OXFORD MEDICAL PUBLICATIONS

Cancer in the Elderly

Cancer in the Elderly

Treatment and Research

Edited by

IAN S. FENTIMAN

ICRF Clinical Oncology Unit,
Guy's Hospital, London

and

SILVIO MONFARDINI

Division of Medical Oncology,
Istituto Nazionale di Ricovero e Cura a Carattere
Scientifico (INRCCS), Italy

OXFORD NEW YORK TOKYO
OXFORD UNIVERSITY PRESS
1994

Oxford University Press, Walton Street, Oxford OX2 6DP

Oxford   New York
Athens   Auckland   Bangkok   Bombay
Calcutta   Cape Town   Dar es Salaam   Delhi
Florence   Hong Kong   Istanbul   Karachi
Kuala Lumpur   Madras   Madrid   Melbourne
Mexico City   Nairobi   Paris   Singapore
Taipei   Tokyo   Toronto
and associated companies in
Berlin   Ibadan

Oxford is a trade mark of Oxford University Press

Published in the United States
by Oxford University Press Inc., New York

© Ian S. Fentiman, Silvio Monfardini, and the contributors listed on pp. ix–xi, 1994

A catalogue record for this book is available from the British Library

Library of Congress Cataloging in Publication Data
Cancer in the elderly : directions for research / edited by Ian S. Fentiman, Silvio
Monfardini.
(Oxford medical publications)
Includes bibliographical references.
1. Geriatric oncology.   I. Fentiman, Ian S.   II. Monfardini, S. (Silvio)   III. Series.
[DNLM: 1. Neoplasms—in old age.   QZ 200 C2151926 1994]
RC281.A34C373 1994   618.97'6994—dc20   94–24864
ISBN 0 19 262200 5

Typeset by The Electronic Book Factory, Fife, Scotland

Printed in Great Britain by
Bookcraft (Bath) Ltd
Midsomer Norton, Avon

Preface

Every year three million people develop cancer. Unless there is a major world catastrophe it is projected that within the next 50 years the number will double. This pandemic is not the result of rises in background irradiation or dietary carcinogens. It will be a consequence of an increase in the numbers of elderly people. Age is the major risk factor for the majority of cancers, half of which become clinically evident in people over 70 years old.

A stereotype has emerged in which age is equated with frailty. Paradoxically, despite the frequency of malignancy in those aged over 70, they have been excluded from the majority of clinical trials. It is from such sources that our decisions on patient management are based. Thus results of studies conducted with younger patients may be applied, possibly wrongly, to an older group. What is more worrying is that because of the attitude of doctors looking after the elderly, such people may be given sub-standard treatments because of a misunderstanding of the nature of the disease and those whom it afflicts.

Problems have arisen partly because of pragmatic decisions related to the design of controlled randomized trials. Investigators were concerned that the effects of treatment might be masked by deaths from intercurrent disease and also that treatment, particularly by chemotherapy, might lead to unacceptable toxicity in the elderly. Additionally it was suspected by some that cancer in the elderly was a more indolent disease.

Careful study of the literature leads inescapably to the conclusion that solid tumours behave neither more nor less aggressively in older patients. Such data as are available confirm that the elderly are often undertreated. Use of suboptimal treatment, whether local or general, will give rise to an increased risk of both recurrence and death from cancer. Seventy-year-old women in Western Europe have a mean life expectancy of 15 years, and men can expect to live to the age of 80. Thus for the majority premature death from degenerative diseases will not mask the effects of inadequate treatment for malignancy.

This is particularly relevant in relation to surgery. With careful preoperative assessment and treatment of coexisting medical conditions such as hypertension, cardiac failure, or diabetes, the majority of patients over 70 can be rendered fit enough to withstand general anaesthetic. This will enable the surgeon to treat effectively the primary tumour. This is almost certainly the only opportunity to cure or control locally the cancer.

We are now beginning to realize that age *per se* tells us very little. What is needed is a system for assessment of frailty which can identify individuals at high risk of death from intercurrent disease. Specially designed trials of treatment for the frail are needed so that optimal therapies can be ascertained.

Most of us will become elderly, that is, live for more than 70 years. A change in medical attitudes is needed. Decisions on treatment of the elderly should take account of the wishes of the individual patient and not just the prejudices of their younger medical attendants. Age biases should be confronted with data derived from prospective randomized trials. Some of the questions which need to be answered are raised in this book. At present we work within a cloud of uncertainty, but sustained clinical and epidemiological effort should enable us to give informed advice to elderly patients with cancer.

London   I. S. F.
Aviano   S. M.
June 1994

Contents

Contributors

Matti S. Aapro

University Hospital, Geneva, and Cancer Center, Clinique de Genolier, 1261 Genolier, Switzerland

Neil K. Aaronson

Department of Clinical Psychology, University of Amsterdam, Roeters-straat 15, 1018 WB Amsterdam, and Division of Psychosocial Research and Epidemiology, The Netherlands Cancer Institute, Plesmanlaan 121, 1066 CX Amsterdam, The Netherlands

S. Ahmedzai

The Leicestershire Hospice, Groby Road, Leicester, LE3 9QE, UK, and The Trent Palliative Care Centre, Sykes House, Abbey Lane, Sheffield, S11 9NE, UK

Vladimir N. Anisimov

Laboratory of Experimental Tumours, N.N. Petrov Research Institute of Oncology, Leningradskaya Str., 68, Pesochny – 2, St Petersburg 189646, Russian Federation

P. Ayela

Department of Hematology–Oncology, Centre Antoine-Lacassagne, 36 Voie Romaine, 06054 Nice Cedex, France

M. Baudard

Service d' Hematologie, Hotel Dieu, 1 Place du Parvis Notre Dame, 75181 Paris Cedex 04, France

Liesbeth Bergman

Division of Psychosocial Research and Epidemiology, The Netherlands Cancer Institute, Plesmanlaan 121, 1066 CX Amsterdam, The Netherlands

A. Creisson

Department of Hematology–Oncology, Centre Antoine-Lacassagne, 36 Voie Romaine, 06054 Nice Cedex, France

Frits S. A. M. van Dam

Department of Clinical Psychology, University of Amsterdam, Roeters-straat 15, 1018 WB Amsterdam, and Division of Psychosocial Research and Epidemiology, The Netherlands Cancer Institute, Plesmanlaan 121, 1066 CX Amsterdam, The Netherlands

I.S. Fentiman

ICRF Clinical Oncology Unit, Guy's Hospital, London, SE1 9RT, UK

A.P.M. Forrest

University Department of Surgery, Royal Infirmary of Edinburgh, Edin-burgh, EH3 9YW, UK, and International Medical College, Kuala Lumpur, Malaysia

M. H. Gaspard

Department of Hematology–Oncology, Centre Antoine-Lacassagne, 36 Voie Romaine, 06054 Nice Cedex, France

N. J. R. George

Department of Urology, Withington Hospital, West Didsbury, Manchester, M20 8LR, UK

Daniela E. E. Hahn

Department of Clinical Psychology, University of Amsterdam, Roeters-straat 15, 1018 WB Amsterdam, and Division of Psychosocial Research and Epidemiology, The Netherlands Cancer Institute, Plesmanlaan 121, 1066 CX Amsterdam, The Netherlands

J. P. Marie

Service d'Hematologie, Hotel Dieu, 1 Place du Parvis Notre Dame, 75181 Paris Cedex 04, France

Silvio Monfardini

Division of Medical Oncology, Centro Regionale di Riferimento Onco-logico, Istituto Nazionale di Ricovero e Cura a Carattere Scientifico (INRCCS), Via Pedemontana Occidentale, 33081 Aviano (PN), Italy

J.M.A. Northover

ICRF Colorectal Cancer Unit, St. Mark's Hospital, City Road, London, EC1V 2PS, UK

J. Otto

Department of Hematology–Oncology, Centre Antoine-Lacassagne, 36 Voie Romaine, 06054 Nice Cedex, France

Michele Pavone-Macaluso

Institute of Clinical Urology, University of Palermo, 90127 Palermo, Italy

P. Scalliet

Department of Radiotherapy, AZ Middelheim, 2020 Antwerp, Belgium

M. Schneider

Department of Hematology–Oncology, Centre Antoine-Lacassagne, 36 Voie Romaine, 06054 Nice Cedex, France

E. van der Schueren

Department of Tumours, University Hospital St. Rafael, 3000 Leuven, Belgium

Vincenzo Serretta

Institute of Clinical Urology, University of Palermo, 90127 Palermo, Italy

A. Thyss

Department of Hematology–Oncology, Centre Antoine-Lacassagne, 36 Voie Romaine, 06054 Nice Cedex, France

Umberto Tirelli

Division of Medical Oncology, Centro di Riferimento Oncologico, Istituto Nazionale di Ricovero e Cura a Carattere Scientifico (INRCCS), Via Pedemontana Occidentale, 33081 Aviano, Italy

D. van den Weyngaert

Department of Radiotherapy, AZ Middelheim, 2020 Antwerp, Belgium

Vittorina Zagonel

Division of Medical Oncology, Centro di Riferimento Oncologico, Istituto Nazionale di Ricovero e Cura a Carattere Scientifico (INRCCS), Via Pedemontana Occidentale, 33081 Aviano, Italy

R. Zittoun

Service d'Hematologie, Hotel Dieu, 1 Place du Parvis Notre Dame, 75181 Paris Cedex 04, France

1

Age-related mechanisms of susceptibility to cancer

Vladimir N. Anisimov

Introduction

Cancer is recognized as a common cause of disability and death in the elderly with over 50 per cent of all cases occurring in those who are over 70.[1,2] However, at present there is no agreement concerning the nature of the relation between ageing and cancer, or concerning causal mechanisms. Peto *et al.* wrote that no relation exists between ageing *per se* and cancer.[3] Subsequently Peto published a paper entitled 'There is no such thing as ageing, and cancer is not related to it.'[4] Cairns has stressed that ageing and cancer are fundamentally different: cancer originates as a clone from a single transformed stem cell, while in ageing the whole organism becomes older.[5] Ageing is a continuous and irreversible process, unlike cancer which has a beginning and may also be reversible.

Pitot, who has kept a question mark in his paper published in 1977 entitled 'Carcinogenesis and ageing—two related phenomena?'[6] recently speculated that 'the ageing process itself exhibits characteristics that favour changes in the stage of progression of neoplasia by endogenous cellular alteration, or that endogenous factors such as dietary energy may promote fortuitous initiated cells'.[7] Dilman proposed 'cancrophilia' as a syndrome, which develops during the process of natural ageing, inevitably causing hormonal and metabolic shifts, facilitating age-related development of pathological conditions including cancer.[8] Dix *et al.*[9] and Ebbesen[10] do not doubt the existence of a close interrelation between ageing and carcinogenesis, while Cutler and Semsei[11] believe that there is indeed considerable evidence indicating common cause of cancer and ageing.

These hypotheses together with extensive data and arguments for and against the interrelation between ageing and carcinogenesis have been reviewed elsewhere.[12,13] The elucidation of causes of an age-related increase in cancer incidence is essential to the elaboration of a strategy for primary cancer prevention.

DNA senescence and carcinogenesis

It has been postulated that both ageing and cancer arise from a common set of genetic alterations.[11] Damage to DNA is implicated in the aetiology of cancer and there is evidence that DNA lesions are major causative factors in ageing.[14,15] One of the hypotheses explaining the interrelation between ageing and cancer development suggests an alteration of methylation in some genes as a possible reason for the increase in cancer in old individuals.[16–18] However, there is no common picture of age-related changes in DNA methylation and patterns of proto-oncogene expression in various tissues.[17–19] Also, the age-related and cancer-associated changes in DNA methylation were not present in all genes which have been tested.[17,20]

The formation of DNA adducts in target tissues in one of the key events in the process in chemical carcinogenesis.[21] Recently it was found that the DNA of various tissues of intact rodents contains adduct-like compounds (I-compounds) which accumulate with age.[22] I-compounds might play an important role in the initiation of both ageing and tumour development. The important characteristics of I-compounds are their capability to cause mutations, DNA chain breaks, and gene rearrangements.[23] However, it is too early to be sure either that these I-compounds are involved in spontaneous tumour development or that they are markers of the ageing process.[24] Accumulation of adduct 7-methylguanine in nuclear and mitochondrial DNA might also play a role in the processes of ageing and carcinogenesis.[25]

Thus, the data available show that some changes in structure and function of DNA are evolving during the process of natural ageing. The character of these changes could vary in different tissues and might cause uneven tissue ageing and thus different patterns of age-related increases in spontaneous tumour incidence in various organs.

Multi-stage model of carcinogenesis and ageing

Because of the positive correlation between the ageing rate of different species and their cancer rates, together with the fact that these processes – ageing and carcinogenesis, may be initiated and promoted by impairments of gene regulation, Cutler and Semsei concluded that both cancer and ageing may arise from a common set of genetic alterations.[11] Some events related to ageing and cancer development have been discussed briefly above. Here the problems will be analysed in terms of the multi-stage model of carcinogenesis. The principles of this model are as follows. Firstly, it is postulated that a normal cell passes through at least one intermediate stage on its way to complete malignant transformation. Secondly, passage from one state to another is a stochastic event, the rate of which depends on the dose of a carcinogen which damages the cell. Finally, all cells at any stage of carcinogenesis may enter the next stage independently of each other.

According to the model, the tumour develops only if at least one cell passes through all the necessary stages, and then as a result of clonal growth forms a clinical cancer. In this model, the exact origin of the various stages is ignored and the changes in cell function during the process of carcinogenesis are not assessed. The grade of malignancy is considered to increase with every stage. Various carcinogenic agents (exogenous as well as endogenous) may be able to modify the rate of multi-stage transformation in different ways. In addition, some agents act at early stages of carcinogenesis and others at later stages.[26] Epidemiological data, analysed within the framework of a multi-stage model, have helped to estimate the value of various factors including the time after the start of carcinogenic exposure, its duration, the time after cessation of carcinogenic exposure, and the age at the onset of exposure.

When carcinogens act at an early stage of neoplastic change, more cells will have passed through all prior stages to become malignant. An increased incidence of tumours induced by such an agent depends on time, and would not decrease immediately on withdrawal of this agent. Carcinogens acting at late stages of carcinogenesis cause the tumour incidence rate to rise without a prolonged latency. The increased rate of tumour incidence will be reversed immediately on cessation of exposure.[26]

The relationship between ageing and carcinogenesis can be described on the basis of results of both epidemiological and experimental studies. There is both a quantifiable time-scale and a qualitative description of events occurring during the passage from normality to malignancy. Figure 1.1 shows an integrated scheme of multi-stage carcinogenesis. At a cellular level carcinogenic agents not only influence the cell, causing genomic transformations that lead to neoplasia, but also create indirectly in the cell microenvironment the conditions that facilitate proliferation and clonal selection.[12]

Multi-stage carcinogenesis is accompanied by disturbances in tissue homeostasis and perturbations in nervous, hormonal, and metabolic factors which may influence antitumour resistance. The time when these changes become manifest depends on the state of those systems at the moment of exposure to a carcinogen or tumour promoter and the dose, and determines the probability of key events in a target stem cell, together

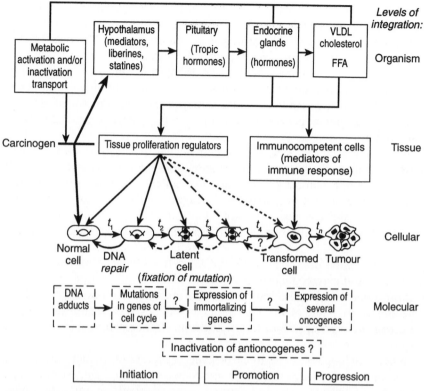

Fig. 1.1 Integral scheme of multi-stage carcinogenesis: $t_1 \ldots t_{ii}$, time of passage of cell from stage k_1 to stage $k_2 \ldots k_n$.

with the probable delay at each intermediate stage for a certain period of time, or even reversion from the malignant pathway. These factors determine the proliferation rate of cells together with the total duration of carcinogenesis and accordingly the latent period of tumour development.

According to the multi-stage model, a carcinogen whose effect increases in proportion to age at exposure is postulated to affect the partially transformed cell. Under these circumstances tumour latency would increase, as compared with a population exposed to the same effective dose of carcinogen at a young age (Fig. 1.2). Numerous experiments support this model. Thus, single skin application with 7,12-dimethylbenz-(a) anthracene (DMBA) in mice aged 8 and 48 weeks at doses ranging from 10 to 30 μg caused increased skin tumour incidence in older mice.[27] Skin application of the promoter 12-0-tetradecanoylphorbol-13-acetate (TPA) in animals of different age caused skin neoplasms only in old animals.[28] Exposure of mice and F344 rats of various ages to phenobarbital resulted in hepatocarcinogenesis only in old animals.[29,30] Single intravenous injection of *N*-nitrosomethylurea (NMU) at doses of 10, 20 and 50 mg/kg was carried out in female rates aged 3 and 15 months.[31] The NMU carcinogenic dose dependence in different age groups was considered in the context of a multi-stage model.

It can be calculated that the number of events necessary for complete malignant transformation in 15-month-old rats under the influence of NMU is less than in 3-month-old rats. It is important to stress that in every tissue the number of events occurring in the stem cell before its complete transformation is variable and depends on many factors, in particular the rate of ageing of the target tissue and its regulatory system(s).[12] This model is consistent with the analysis of age-related distribution of tumour incidence in different sites in humans and experimental animals.[2,12]

Life-span prolongation and carcinogenesis

Among the approaches to studying the mechanisms responsible for the interrelation between ageing and cancer, special attention must be given to the relationship between factors or drugs that increase life-span (geroprotectors) and the rate of tumour development. Available data and results of our studies on spontaneous tumour incidence in animals exposed

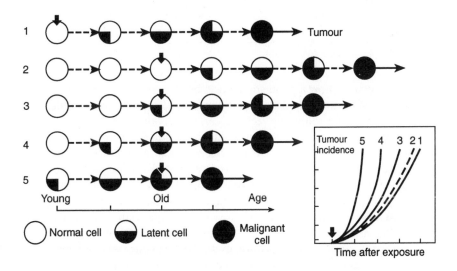

Fig. 1.2 The multi-stage carcinogenesis induced by single exposure to a carcinogenic agent at different ages. Groups 3–5 demonstrate the carcinogenic effect produced on a cell that has passed through one or more stages in accordance with the multi-stage model of carcinogenesis.

Table 1.1 Effects of life-span prolonging drugs (geroprotectors) on spontaneous carcinogenesis in rodents[a]

Type of delay ageing	Geroprotector	Effect on	
		Tumour latency	Tumour incidence
I	2-mercaptoethylamin	Increases	No effect
	2-ethyl-6-methyl-3-oxipyridine	Increases	No effect
	Procaine (gerovital)	No effect	No effect
II	Caloric-restricted diet	Increases	Decreases
	Tryptophan-deficient diet	No data	Decreases
	Phenoformin, buformin	Increases	Decreases
	L-DOPA	No effect	Decreases
	Diphenylhydantoin	No effect	Decreases
	Dehydroepiandrosterone	Increases	Decreases
	Succinic acid	No effect	Decreases
	Epithalamin (pineal peptide preparation)	Increases	Decreases
	Thymalin (thymic peptide)	Increases	Decreases
	Levamisole	Increases	Decreases
III	Selenium	No data	Increases
	EDTA-sodium	No data	Increases
	Tritium oxide (low doses)	No data	Increases
	Tocopherol (vitamin E)		
	benign tumours	Increases	Increases
	malignant tumours	Increases	Decreases

[a] Full bibliography given in refs. 12 and 13.

to geroprotectors show a good correlation between the type of ageing delay and pattern of tumour development induced by geroprotectors (Table 1.1).

Thus, life-span prolonging agents can be divided into three groups: (a) geroprotectors that prolong the life-span equally in all members of the population—these postpone the beginning of population ageing; (b) geroprotectors that decrease the mortality of a long-living subpopulation leading to a rise in maximal life-span—these slow down the population ageing rate; (c) geroprotectors that increase the survival in a short-living subpopulation without a change in the maximal life-span—in this case the ageing rate increases (Fig. 1.3). As can be seen in Table 1.1 and Fig. 1.3, geroprotectors of the first type do not influence the incidence of tumours but do prolong tumour latency. Geroprotectors of the second type are effective in inhibiting carcinogenesis, prolonging tumour latency, and decreasing the incidence of cancers. Drugs of the third type can sometimes increase tumour incidence in exposed population.[12,13]

The mean life-span of animals increases equally in those given the three types of geroprotectors (Fig. 3), revealing no correlation between these life-span prolonging agents and tumour rate in the population. At the same time, the correlation between the ageing rate (the ageing rate can be estimated as α in the Gompertz equation $R = R_0 e^{\alpha t}$, where R is mortality, $R_0 = R$ at time $t = 0$, and α is a constant) and malignant tumour incidence in the populations appeared to be high.[12] Within the framework of the multi-stage model, geroprotectors of one type or another may either inhibit or accelerate the passage of cells exposed to a carcinogen from one stage to another. Under these circumstances the efficacy of geroprotectors in preventing cancer development would decrease in relation to the age at onset of exposure. It is important to stress that geroprotectors of the second type that slow down the ageing rate do this by influencing the 'main' regulatory systems of the organism (nervous, endocrine, immune). Thus, there is a slow-down of age-related changes in the microenvironment of cells exposed to carcinogenic agents or stochastic hits.

Geroprotectors may also be classified into two main groups according to their mechanism of action. The first group includes those drugs that mainly prevent stochastic damage to biological macromolecules. The theoretical basis for using these drugs is provided for by variants of the 'catastrophe

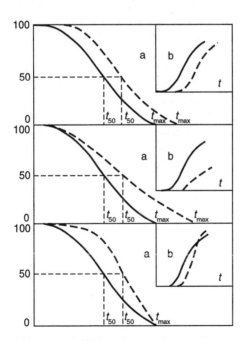

Fig. 1.3 (a) Types of ageing delay (ordinate, number of survival animals (per cent); abscissa, age), and (b) incidence of spontaneous tumours (ordinate, tumour rate (per cent); abscissa, age) under the influence of geroprotectors: — control,— — administration of geroprotector.

error' theory, which regards ageing as a result of the accumulation of stochastic lesions. The second group includes drugs and factors which are thought to slow down the programme of ageing and the development of age-related pathology.

Antioxidants are the most typical representatives of the first group. Their geroprotective and antitumour effects depend on the age at first administration and depend inversely on the dose of the damaging agent. The second group is represented by the antidiabetic biguanides (phenformin and buformin), epithalamin (pineal peptide factor), and caloric-restricted diets. These factors influence the hormonal, metabolic, and immunological patterns in the body, leading to normalization of their age-related shifts and, thus, providing their antitumour effects.[12]

The data presented could provide an explanation for the currently observed age-related increased incidence of cancer. The survival characteristics of human population were noted to become more and more 'rectangular'.[32] This is caused mainly by the decrease in child and early mortality, which is connected with infectious and some non-infectious diseases. As a result, a significant increase in the mean life-span in human populations occurred, while the maximum life-span has stayed the same since ancient times.[33] The changes in shapes of the survival curves of human populations correspond to the third type of ageing delay presented in Fig. 3. The changes of this type were shown experimentally and epidemiologically to be associated with an increase in tumour rate. In other words, mankind pays for the increase in mean life-span achieved by the decrease in mortality of the young, by an increased risk of cancer and some other diseases of civilization such as atherosclerosis or diabetes mellitus at old age.

Strategies for cancer prevention must include not only measures to minimize exposure to exogenous carcinogenic agents, but also measures to normalize the age-related changes in internal millieu and thus slow down the genetic programme of ageing. In conclusion, it is useful to remember E.S. Bauer[34] who wrote 58 years ago; 'The problem of cancer apparently coincides with the problem of senility. The aim of science is to slow down the process of ageing and in this way it may reduce the probability of cancer development'.

Key points

- Although there is a relation between cancer and ageing, the two problems are independent.
- There is no clear picture of alteration in DNA methylation and age.
- Carcinogenesis is a multi-stage process with quantifiable time-scale and identifiable events occurring during the process.
- Carcinogenesis with age-related effects may affect partially transformed cells with a decrease in tumour latency.

- Geroprotectors may postpone population ageing and prolong tumour latency. Alternatively they can decrease the mortality in long-lived populations and inhibit carcinogenesis. Finally, geroprotection may increase the length of survival of a short-lived population but increase the incidence of malignancy.

References

1. Napalkov, N.P. (1985). Relation of human cancer morbidity to age: general patterns and exceptions in the USSR. In *Age related factors in carcinogenesis*, IARC Sci. Publ. No. 58, (ed. A. Likhachev, V. Anisimov, and R. Montesano), pp. 9–20. IARC, Lyon.
2. Dix, D. (1989). The role of ageing in cancer incidence: an epidemiological study. *J. Gerontol.*, **44** (6) 10–18.
3. Peto, R., Roe, F.J.C., Lee, P.N., Levy, L., and Clack, J. (1975). Cancer and aging in mice and men. *Br. J. Cancer*, **32**, 411–26.
4. Peto, R., Parish, S.E., and Gray, R.G. (1985). There is no such thing as aging, and cancer is not related to it. In *Age related factors in carcinogenesis*, IARC Sci. Publ. No. 58, (Ed. A. Likhachev, V. Anisimov, and R. Montesano), pp. 43–53. IARC, Lyons.
5. Cairns, J. (1982). Aging and cancer as genetic phenomena. *Natl. Cancer Inst. Monogr.*, **60**, 237–9.
6. Pitot, H.C. (1977). Carcinogenesis and aging—two related phenomena? *Am. J. Pathol.*, **87**, 444–72.
7. Pitot, H.C. (1989). Aging and cancer: some general thoughts. *J. Gerontol.*, **44** (6), 5–9.
8. Dilman, V.M. (1985). Aging and cancer in the light of ontogenetic 'model of medicine'. In *Age related factors in carcinogenesis*, IARC Sci. Publ. No. 58, (Ed. A. Likhachev, V. Anisimov, and R. Montesano), pp. 21–3. IARC, Lyons.
9. Dix, D., Cohen, P., and Flannery, J. (1980). On the role of aging in cancer incidence. *J. Theor. Biol.*, **83**, 163–73.
10. Ebbesen, P. (1984). Cancer and normal aging. *Mech. Ageing Dev.*, **25** 269–83.
11. Cutler, R.G. and Semsei, I. (1989). Development, cancer and aging: possible common mechanisms of action and regulation. *J. Gerontol.* **44** (6), 25–34.
12. Anisimov, V.N. (1987). *Carcinogenesis and aging*, Vols. 1 and 2. CRC Press, Boca Raton, FL.
13. Anisimov, V.N. (1989). Age-related mechanisms of susceptibility to carcinogenesis. *Semin. Oncol.* **16**, 10–19.
14. Kirkwood, T.B.L. (1989). DNA, mutation and aging. *Mutat. Res.*, **219**, 1–7.
15. Strehler, B.L. (1986). Genetic instability as the primary cause of human aging. *Exp. Gerontol.* **21**, 283–319.
16. Nyce, J., Weinhouse, S., and Magee, P. (1985). 5-Methylcytosine depletion during tumour development: an extension of the miscoding concept. *Br. J. Cancer*, **48**, 463–75.
17. Ono, T., Takahashi, N., and Okada, S. (1989). Age-associated changes in DNA methylation and mRNA level of the c-myc gene in spleen and liver of mice. *Mutat. Res.*, **219**, 39–50.

18. Mays-Hoopes, L.L. (1989). Age-related changes in DNA methylation: do they represent continued developmental changes? *Int. Rev. Cytol.* **114**, 181–220.
19. Semsei, I., Ma, S., and Cutler, R.G. (1989). Tissue and age specific expression of the myc proto-oncogene family throughout the life span of the C57BL/6J mouse strain. *Oncogene*, **4** 465–70.
20. Ono, T. (1990). Changes of DNA methylation in aging and carcinogenesis. *J. Cancer Res. Clin. Oncol.* **116** (Suppl.), 1056.
21. Singer, B. and Gruneberger, D. (1983). *Molecular biology of mutagens and carcinogens*. Plenum Press, New York.
22. Randerath, K., Liehr, J.G., Gladek, A., and Randerath, E. (1989). Age-dependent covalent DNA alterations (I-compounds) in rodent tissues: species, tissue and sex specification. *Mutat. Res.*, **219**, 121–33.
23. Randerath, K., Reddy, M.V., and Disher, R.M. (1986). Age and tissue-related DNA modifications in untreated rats: detection by ^{32}P-post-labelling assay and possible significance for spontaneous tumour induction and aging. *Carcinogenesis*, **7**, 1615–17.
24. Warner, H.R. and Price, A.R. (1989). Involvement of DNA repair in cancer and aging. *J. Gerontol.*, **44** (6), 45–54.
25. Park, J.W. and Ames, B.N. (1988). 7-Methylguanine adducts in DNA are normally present at high levels and increase on aging: analysis by HPLC with electrochemical detection. *Proc. Natl. Acad. Sci. USA*, **85**, 7467–70.
26. Kaldor, J.M. and Day, N.E. (1987). Interpretation of epidemiological studies in the context of the multistage model of carcinogenesis. In *Mechanisms of environmental Carcinogenesis*, Vol. 2, (ed. J.C. Barret), pp. 21–57. CRC Press, Boca Raton, FL.
27. Stenback, F., Peto, R., and Shubik, P. (1981). Initiation and promotion at different age and doses in 2200 mice: III linear extrapolation from high doses may underestimate low-dose tumour risk. *Br. J. Cancer*, **44**, 23–34.
28. Ebbesen, P. (1985). Papilloma development on young and senescent mouse skin treated with 12-0-tetradecanoylphorbol-13-acetate. In *Age-related factors in carcinogenesis*, IARC Sci. Publ. No. 58, (ed. A. Likhachev, V. Anisimov, and R. Montesano), pp. 167–70. IARC, Lyons.
29. Ward, J.M. (1983). Increased susceptibility of livers of aged F344/NCr rats to the effects of phenobarbital on the incidence, morphology, and histochemistry of hepatocellular foci and neoplsms. *J. Natl. Cancer Inst.*, **71**, 815–23.
30. Ward, J.M., Lynch, P., and Riggs, C. (1988). Rapid development of hepatocellular neoplasms in aging male C3H/HeNcr mice given phenobarbital. *Cancer Lett.*, **39**, 9–18.
31. Anisimov, V.N. (1988). Effect of age on dose–response relationship in carcinogenesis induced by single administration of *N*-nitrosomethylurea in female rats. *J. Cancer Res. Clin. Oncol.*, **114**, 628–35.
32. Hirsch, H.R. (1982). Evolution of senescence: natural increase of population displaying Gompertz or power-law death rates and constant of age-dependent maternity rates. *J. Theor. Biol.*, **98**, 321–46.
33. Gavrilov, L.A. and Gavrilova, N.S. (1986). *The biology of life span: quantitative aspects*. Nauka, Moscow.
34. Bauer, E.S. (1936). Cancer as a biological problem. In *Modern problems of theoretical medicine*, Vol. 1, (ed. R.E. Jackson) pp. 37–45. State Publ. House Biol. Med. Lit., Moscow.

2

Psychosocial issues

Daniela E.E. Hahn, Liesbeth Bergman,
Frits S.A.M. van Dam, and Neil K. Aaronson

Cancer is primarily a disease of the elderly. The risk of developing cancer doubles every 5 years after the age of 25. More than half of all cases, particularly cancer of the breast, colon, lung, and prostate, occur in persons 65 years of age or older. Sixty per cent of all cancer mortality is reported in this age group (Frank-Stromborg 1988; Given and Given 1989; Kennedy 1991; Kleinpell and Foreman 1990; Yancik and Ries 1989). The combination of ageing and risk of cancer forms an increasingly prevalent health problem. Despite the magnitude of this health problem, there is a paucity of descriptive and controlled randomized clinical studies on the optimal medical and psychosocial management of the elderly cancer patient. The major sources of information concerning suitable treatment regimens are clinical trials of cancer therapy. Unfortunately, these studies often include an age limit and are carried out on patients younger than 70 years of age (Fentiman *et al.* 1990; Yancik 1983; Yancik and Ries 1989).

Disease unquestionably affects 'quality of life', defined as the physical, psychological, and social well-being of the individual. Recently, attention has been drawn to the possible impact of medical treatment for cancer on 'quality of life' (Aaronson 1988, 1991; Spilker, 1990). In this chapter, the extant literature on psychosocial issues related to cancer in elderly people is reviewed. The different stages in the cancer illness trajectory are used as a framework for a description of the psychosocial areas where comprehensive care for the elderly cancer patient is necessary. Firstly, the issue of early tumour detection is addressed. This includes responses of the elderly to signs and symptoms of a serious disease and possible delay in seeking medical care. Secondly, studies on treatment motives and choices for elderly cancer patients are reviewed. Emphasis is placed on aspects of quality of life for the aged chronically ill. Complications in the assessment of quality of life and the impact of treatment choices on the quality of life of elderly cancer patients are discussed. Thirdly, the results of psychosocial studies on cancer in the elderly are reviewed briefly. The chapter closes with recommendations for targeted psychosocial research concerning elderly cancer patients, both in detailed descriptive studies and within controlled randomized prospective clinical trials.

The diagnostic process

Early diagnosis of cancer is as important for elderly as for younger people. The elderly are a considerable and distinctive target group for the prevention and early detection of neoplasms. Early detection may result in reduced morbidity, longer disease-free periods following initial treatment, and reduced mortality (Warnecke 1989). Unfortunately, people of 65 years and older are seldom the target of screening and other health promotion campaigns. Symptom-detection campaigns among the elderly are needed because of increased difficulties of the elderly in perceiving quickly the alarming signs and symptoms of cancer (Ouslander and Beck 1982). Some neoplasms have a longer latency period in the elderly, resulting in the gradual emergence of symptoms in the later years. Potentially, symptoms may be masked by limitations in physiological function and concomitant diseases common at older age. Elderly people, as well as their care-givers and primary health care providers, need to be made aware of the fact that they may exhibit age-specific manifestations of disease and tend toward under-reporting of signs and symptoms of cancer.

Time of diagnosis

Recent studies disclose a tendency for certain tumours to be diagnosed at a more advanced stage of disease in older people. However, the literature on the relationship between age and disease-stage at time of diagnosis is somewhat inconsistent and findings vary by cancer site (Bergman *et al.* 1991a; Goodwin *et al.* 1986; Holmes and Hearne 1981; Mor *et al.* 1989). Data from population-based studies indicate that cancer of the bladder, breast, cervix, ovary, uterus, and melanoma tend to be diagnosed at a more advanced stage in the elderly, as compared with younger patients (Goodwin *et al.* 1986; Holmes and Hearne 1981). For patients with colon tumours, no relationship was found between age and stage of tumour. For those with bronchial, rectal, and gastric cancers, Goodwin *et al.* noted a significant inverse association, whereas Holmes and Hearne found no significant relationship for cancers of rectum and stomach. The study of Mor *et al.* (1989) revealed no age effect for extent of disease at diagnosis for cancers of the breast, lung, and colorectum. Goodwin and others (1986) found an inverse relationship between (older) age and tumour stage at diagnosis for both lung and gastrointestinal cancers. One of the possible explanations provided was incomplete staging. Especially for lung and gastrointestinal cancers, complete staging often requires elaborate surgical procedures which physicians may avoid when dealing with elderly patients. Less extensive staging procedures for elderly patients may result in classification into a lower stage group than that which corresponds to the true extent of disease, the so called 'Will Rogers' phenomenon (Feinstein *et al*. 1985). This falsely favourable staging may lead, in turn, to the choice of less intensive treatment regimens. As a result, response rate and stage-related survival may be more easily misinterpreted in elderly patients.

Patient delay

'Patient delay' is mentioned frequently as a key reason for presenting with a more advanced stage of disease at the time of diagnosis. Psychosocial factors associated with patient delay in seeking treatment are far from being fully understood, especially for the elderly (Hahn and van Dam 1988). Several authors report a reluctance among elderly patients to undergo screening activities such as cervical cytology or breast self-examination (Hobbs *et al*. 1980; Kegeles and Grady 1982; Warnecke 1989; Yancik and Ries 1989). The process of seeking medical care in symptomatic patients involves a complex interaction of factors (Berkanovic and Aaronson 1986; Cacioppo *et al*. 1986, 1989; Safer *et al*. 1979). At a symptomatic level, several key, related factors are (a) (lack of) motivation to understand unexplained bodily changes, (b) (lack of) knowledge of (alarming) signs and symptoms, and (c) the (in-)ability to disentangle cancer-related symptoms from those associated with the 'normal' process of ageing or with concurrent diseases (Frank-Stromborg 1986; Given and Given 1989; Keller *et al*. 1988; Leventhal and Prohaska 1986; Leventhal *et al*. 1988; Ouslander and Beck 1982; Prohaska *et al*. 1987; Warnecke 1989; Weinrich and Weinrich 1986). Symptom characteristics and restrictions in daily life due to symptoms appear to be strong predictors of the decision to seek medical care promptly. Once a symptom has been acknowledged, negative feelings and concerns about the future may delay further the decision to see the doctor (Love 1991; Nylenna 1986; Riley 1983; Rimer 1983).

The elderly may be especially concerned about the possibility of losing their independence. This may lead to the tendency to under-report health complaints. Also, older patients may be reluctant to submit themselves to extensive diagnostic procedures. Delay in seeking medical attention is affected further by misconceptions about the impact of treatment and by more general beliefs about health and illness (Warnecke 1989; Wolinsky *et al*. 1988). Both access to health care and the quality of the patient–doctor relationship have also been shown to influence the length of time which elapses between the onset of symptoms and seeking a medical diagnosis (Safer *et al.* 1979; Timko 1987).

The role of social support in the decision to seek medical attention for suspicious symptoms has been investigated in several studies. Kahn and Antonucci (1984) conducted a prospective study on reasons for seeking medical care and predictors of patient delay among 718 elderly patients. The advice of social network members was found to be a strong predictor of the *decision* to see a doctor. However, social network influences failed consistently and significantly to predict delay in consulting the doctor. In their study of 1009 individuals aged 65 years or older, Berkanovic and his coworkers (1988) were unable to demonstrate any association between psychological distress or social support and *seeking* medical care. Similarly, Hahn and her coworkers (1990) found no significant association between social support and patient delay in their study of elderly patients with gynaecological, urological, and colorectal cancers. The tendencies to dismiss potentially significant symptoms as minor, to ascribe symptoms to a concomitant disease, and to view them as a reflection of the ageing process, were found to be the most consistent predictors of symptom perception and delay.

These findings suggest the need for interventions focused on improving accurate symptom perception among the elderly. Such approaches could take the form of a special education programme on those cancer signs and symptoms that are easily misattributed to the process of ageing or to other concurrent chronic diseases. In a similar manner as in the public awareness campaign of the American Cancer Society (1990), such a health education programme might focus on 'the seven warning signs of cancer'. Health messages could be delivered by special health education programmes oriented towards the elderly (Timer 1986), via the mass media, or could be integrated into existing social and health services in the community (Borgers and Visser 1990; Plagge 1990; Warnecke 1989). Family members of the elderly form another important target group. Besides monitoring changes in appearance and functioning of the elderly, family members need to encourage and actually to prompt the seeking of medical care. Because the elderly are often afraid of becoming a burden to their helpers or being considered hypochrondriacs and 'doctor shoppers' they tend to under-report symptoms of illness (Ouslander and Beck 1982; Warnecke 1989; Winograd 1986). Therefore, the importance of reporting any concern about health issues needs to be stressed.

Doctors' delay

Late detection of cancer in the elderly may also be caused by physicians. In just the same way as patients, doctors may mistake the symptoms of disease for signs of ageing or misattribute symptoms and a declining functional status to other aetiologies in the elderly (Goldsmith and Brodwick 1989; Leventhal 1984; Mold *et al.* 1991). For instance, constipation heralding colonic carcinoma may be (mis-)attributed to difficulties common in later life (Winograd 1986). Physicians may also be reluctant to carry out adequate screening procedures. Health care providers may perceive the elderly as being too frail to undergo intensive diagnostic procedures. Misconceptions about the capabilities of the elderly effectively to learn self-examination practices may also lead health care providers to avoid promoting special education programmes such as breast self-examination (Adelman *et al.* 1987; Borgers and Visser 1990; Greene *et al.* 1987). Medical education programmes for the elderly should focus on optimizing general practitioners' awareness of masked symptoms of cancer in the elderly, the potential for under-reporting of symptoms, and the benefit of intensive diagnostic procedures.

The therapeutic process

The treatment biases of health care professionals may influence cancer treatment for the elderly (Given and Given 1989; Greenfield *et al.* 1987; Kennedy 1991; Samet *et al.* 1986; Yancik 1989; Yancik and Ries 1989). Optimal oncological care aims at balancing probabilities of cure and toxicity, and chance of palliation with quality of life (Fentiman *et al.* 1990). Elderly patients should not be an exception to this practice. Patterns of care using a less aggressive approach in the elderly may occur because of assumptions and biases of health care providers, rather than because of hard clinical data regarding the effectiveness of treatment and toxicity.

Age by itself is at best a crude indicator of health and certainly should not guide treatment choices. It is more important that physicians determine the actual level of functioning of the elderly patient before selecting the most appropriate treatment plan (Fentiman *et al.* 1990; Kane 1983; Yancik 1989). A priori there is no reason for employing different therapeutic goals for younger patients and elderly patients. Why should the achievement of a long-lasting complete remission be the therapeutic goal for a person of 50 years whereas in elderly patients the effective stabilization of the disease with an acceptable quality of life is viewed as sufficient? One should not underestimate the life expectancy of the elderly, or rule out *potentially* curative treatment options because of an uncritical reference to elderly cancer patients' supposed frailty or an impaired quality of life (Hahn and van Dam 1988). A gradually declining health status with increasing age and concomitant chronic diseases are clinical problems to be considered seriously in oncological medical decision-making. Only by

careful consideration of all medical and psychosocial aspects of cancer
and its treatment in elderly patients and by estimating accurately patients'
resources will the most suitable treatment strategies be found. As quality
of life considerations emerge as legitimate evaluation parameters in medical
decision-making, more attention should be focused on the measurement
and integration of quality of life aspects which may be of special concern
to the elderly.

Treatment choices and motives

There is a growing appreciation among the general public as well as in
the health professions of the need to incorporate patients' preferences
into medical decision-making (McNeil *et al.* 1988). Significant age-trends
in attitudes toward information and participation preferences in medical
decisions have been found in studies carried out by Cassileth *et al.* (1980),
Derdiarian (1987), and Menko (1986). In general, these studies suggest that
the older the patients, the more likely they are to conform to a traditional,
non-participatory patient role. Decision-making under conditions of mental
decline is a particularly relevant issue in the care of the elderly, owing to
the relatively high prevalence of dementia, delirium, and cognitive decline
(Holland and Massie 1987; Pearlman and Uhlmann 1988).

Little research has been conducted on treatment motives among elderly
cancer patients. A recent study indicates that decisions concerning the
choice of treatment are influenced by many factors. Some medical decisions have a sound scientific basis (low health status, serious concomitant
diseases) while others are based on highly personal opinions of physicians,
and on the type of hospital in which medical care is being provided (Givio
1986). Treatment motives are not always well documented in medical
records. Physicians may record age as the reason for deviating from
standard treatment, whereas actually the physical condition of the patient
is the primary consideration (Bergman *et al.* 1991). Conversely, Silliman *et
al.* (1989) noted that age often plays an important role in treatment choice,
although this is frequently left out of the medical record.

The roles of age *per se* and comorbidity as motives for treatment choice
in the elderly are not yet entirely clear. Greenfield and his coworkers
(1987) report that elderly breast cancer patients with a good health
status received less intensive treatment than younger patients. Mor *et
al.* (1989) and Bergman *et al.* (1991) concluded from a medical records
audit that age *per se*, rather than comorbidity, often influences specialists'
treatment choices.

Age bias in treatment planning has been confirmed by several recent
studies. For example, elderly breast cancer patients have been found to be

treated less intensively than younger patients (Allen *et al.* 1986; Bergman *et al.* 1991; Chu *et al.* 1987; Donegan 1983; Greenfield *et al.* 1987; Mor *et al.* 1985; Samet *et al.* 1986). Following mastectomy, elderly patients (over the age of 75 years) receive less adjuvant radiotherapy than younger patients. Primary hormonal therapy is also used more often as the only treatment for elderly patients with operable disease.

There is a pressing need to convince both general practitioners and specialists that all treatment options should be considered, when indicated, irrespective of the age of the patients (Tirelli *et al.* 1991). In our view, the patient's physical condition, prior health status, the presence of comorbidity rather than age *per se*, and quality of life aspects should be used in medical decision making. Feinstein and coworkers (1985) have developed a clinical severity staging system, classifying the clinical condition of cancer patients according to symptom pattern, symptom severity, and severity of comorbidity. Another, more traditional, prognostic index based on the tumour characteristics is the TNM system. The use of Feinstein's classification system in combination with the more traditional TNM could significantly reduce the inappropriate recourse to patients' age as an overriding factor in medical decision-making.

This combination must be supplemented by information concerning the patients' preferences. Regardless of age, patients should be encouraged to express their attitudes, wishes, and concerns regarding treatment options. Objective guidelines, expert opinion, and patients' preferences should all exert influence on the treatment planning process.

Quality of life

Quality of life has been increasingly recognized as an important component of health status assessment and treatment evaluation in oncology (Aaronson and Beckmann 1987; Osoba 1991; Spilker 1990). While the assessment of tumour response, disease-free survival, and overall survival remains central in clinical research, quality of life assessment can be used for the following

(1) to describe the functional and psychosocial problems confronting patients;
(2) to establish norms for psychosocial complications among specific patient groups;
(3) to screen for behavioural and/or psychopharmacological interventions;
(4) to control the quality of care;
(5) to evaluate the efficacy of competing medical or psychosocial interventions.

A universal definition of the quality of life concept remains elusive. Most researchers agree that quality of life is a multidimensional construct composed minimally of the following four domains: (1) physical and role functioning; (2) disease-related and treatment-related signs and symptoms; (3) psychological functioning; (4) social functioning. While oncologists may focus on disease-related outcomes, for instance tumour response, patients may be equally concerned with the consequences of cancer and its treatment for their daily life. In this sense, quality of life data can contribute to translating medical outcomes into terms that are easily understood by the patient and the patient's family (Aaronson 1988).

All four quality of life domains can be affected by cancer and cancer treatment, regardless of the age of the patient (Williams 1990). Physical and role functioning refers to the ability to perform a range of daily activities that are normal for most people of a given age, including the categories of self-care, mobility, physical activities, and (social) role activities. The physical dimension of quality of life may be particularly important to elderly cancer patients in that a decrease in functional ability may result in a loss of independence. Disease-related and treatment-related symptoms refer to the range of physical symptoms reported frequently by cancer patients, either as a result of the disease process or of the antitumour treatment. Psychological functioning refers to both levels of psychological distress and levels of well-being. Careful consideration should be given to the assessment of cognitive decline in the elderly and psychological disorders resulting from certain multidrug systemic treatments or certain combinations of anticancer and other drug therapies used to treat comorbid conditions (Holland and Massie 1987). Social functioning refers to the level of social activities that are common for patients once they have been diagnosed as having cancer and in the subsequent phases of the illness. Disruption of relationships can result from embarrassment, the fear of being a burden to others, or from restrictions in functional capacities (Aaronson 1988).

Assessment of quality of life in the elderly

There is need for accurate assessment of the complicated interaction of medical, psychological, and social factors in elderly cancer patients. Comparative studies of the relative importance attached to various quality of life domains by the elderly and by younger patients would be valuable.

Quality of life measures for elderly patients need to be sensitive enough to measure (temporary) restrictions in daily function and separate the impact of disease symptoms and treatment-related side-effects from the normal process of ageing. Functional status measures should be extended to assess the family member resources and support needed for aggressive treatment in times when these resources may be compromised (Given and Given 1989; Foreman and Kleinpell 1990).

It is widely held that the patient is the most appropriate source of information on his or her quality of life. Thus quality of life measures are most often based on patients' self reports (e.g. interviews or questionnaires). Nevertheless, there remain selected situations in which quality of life ratings cannot be obtained from the patient himself or herself, owing to severity of symptoms or cognitive deficit. Such situations may arise more frequently with elderly cancer patients. Alternative sources of information on the quality of life of elderly patients include the physician, nurses, and family members. While such individuals may be able to provide important insights into the well-being and functioning of elderly patients, such 'proxy' ratings should nevertheless be interpreted with a good deal of caution (Aaronson 1991; Sprangers and Aaronson 1992).

In assessing the quality of life of the elderly cancer patient, special attention should be devoted to the length and format of the measures employed (Borgers and Visser 1990; Foreman and Kleinpell 1990; Warnecke 1989; Yancik and Ries 1988). Especially for elderly persons with poorer health status and declining visual and auditive capacities, quality of life measures need to be brief and simple to complete.

Psychosocial consequences of cancer treatment

Essential features of quality of life in old age and their interrelations with quality of care are well described by Williams (1990) and Yancik and Ries (1989). Contrary to pervasive earlier views of inevitable functional and mental decline, changes in personality, and loss of social involvement with increasing age, recent (gerontological) research reports on the stability of these quality of life aspects for individuals in their 80s or older, in the absence of a chronic disease. Mages and Mendelson (1981) point to an accelerated psychological disengagement of elderly patients after the diagnosis of cancer. Others, however, highlight the stable (psychological) adaptation resources of elderly cancer patients (Massie and Holland 1987; Nerenz *et al.* 1986; Williams 1990).

Descriptive studies on the psychosocial consequences of cancer treatment which compare older and younger individuals yield surprisingly consistent results. The results from key studies on treatment adaptation reveal no greater general psychological problems among elderly than among younger patients. Nerenz and his colleagues (1986) investigated the presence, duration, and severity of side-effects, emotional adaptation, and difficulties due to treatment in 217 patients (age 19–83 years) receiving initial chemotherapy treatment for breast cancer or lymphoma. Interviews designed to monitor psychosocial functioning were carried out repeatedly over the first six months of treatment. Fewer side-effects from treatment were reported by the age group 70–83 years than by the younger age groups. In the same study, a significantly higher incidence of anticipatory nausea or vomiting was found among younger patients. According to the authors, elderly patients adjust more easily to chemotherapy because they have learned how to live with the problems of illness.

Ganz and her colleagues (1985) found no significant differences within a group of 240 male cancer patients of varying ages (different tumour sites) regarding communication with the medical team, therapeutic regimen-related problems, or compliance and discomfort during the medical procedure. However, younger patients (below 65 years) experienced more frequent and more severe psychosocial problems than older patients. Maisiak *et al.* (1983) reported that cancer patients aged 60 years and older had a better psychosocial status, expressed fewer signs of depression, and managed work and leisure better than the younger patient group. However, because of differences in therapy regimens (Nerenz study), the diversity of cancer diagnoses, the exclusion of female patients, and the lack of control for tumour stage (Ganz study) these results need to be viewed with caution. Mor and his colleagues (1989) investigated the influence of age, performance status, and medical and psychosocial status on the management of 700 cancer patients (different cancer sites, local and metastatic disease). They found that the evaluation of quality of life and mood did not vary as a function of age.

Taken together, these studies suggest that elderly patients can tolerate cancer therapies, even when such therapies are relatively aggressive in nature. The widely held belief that elderly patients are too frail to undergo certain forms and intensities of treatment does not seem warranted. However, as most of the research in this area has been correlational in nature, there is need for experimental, methodologically sophisticated studies of the role of age in treatment options, planning, and policy.

Compliance

Non-compliance with medical prescriptions and treatment is a significant problem in many areas of health care (Haynes *et al.* 1976). Older cancer patients have been found to be more likely to comply with therapeutic regimens than younger patients (Haug 1979; Massie and Holland 1989). However, there is also evidence from the gerontology literature that social isolation, depressive mood state, and cognitive dysfunction in the elderly are related to maladaptive health behaviour and non-compliance with therapeutic regimens. Levy (1983) reports that, as the complexity of treatment regimens increases, so does non-compliance with the regimen. For older people with (multiple) chronic illness, the perceived complexity of regimen requirements may be amplified either by memory lapses or by depression and social isolation. Bonnadonna and Valagussa (1981) found that elderly breast cancer patients did not follow the fully prescribed adjuvant chemotherapy.

Even if regimens are not complex, as is the case with the primary treatment of elderly breast cancer patients with tamoxifen, non-compliance may be a serious problem. This treatment requires more frequent follow-up visits which may be experienced by the elderly as burdensome (Bergman *et al.* 1992). This might be used as an argument in favour of primary surgical treatment above primary hormonal treatment in elderly breast cancer patients with operable disease. Similarly, scheduling radiotherapy may prove difficult for elderly patients who live in homes for the elderly or who depend on others for their transportation. Too great a distance from home to the hospital has been found to affect adversely patients' adherence to cancer treatment and to follow-up scheduling. From a psychosocial point of view, the presence of supportive family members or staff members in nursing homes who encourage adherence to medical treatment may be crucial for the elderly patient (Foreman and Kleinpell 1990).

Conclusions and recommendations

As long as the current practice of excluding the elderly from controlled clinical trials remains unchallenged, elderly patients are at risk of receiving either untested treatments, inadequate treatments, or even no treatment at all (Fentiman *et al.* 1990). Early symptom detection and the determination of optimal cancer treatment for the elderly requires cooperative efforts among oncologists, geriatricians, pharmacologists, and psychologists. Further refinements in adequate psychosocial management of the elderly cancer patient require the following types of research:

(1) descriptive and controlled prospective studies of the obstacles to early cancer detection among the elderly;

(2) assessment of functional complications of cancer and its treatment in the elderly;

(3) quality of life assessment in descriptive and randomized prospective trials without upper age-limits;

(4) development, implementation, and evaluation of health education programmes for the elderly and for health professionals;

(5) studies of elderly patients' management needs during active treatment, as well as in palliative and terminal care phases;

(6) controlled studies on compliance with therapeutic regimens in elderly cancer patients.

From a methodological perspective, future studies should focus on improving comprehensive generic and cancer-specific quality of life questionnaires and on developing specific instruments for elderly cancer patients. Such instruments should be sensitive to small but clinically significant changes in health status and should be able to differentiate between the impact of disease symptoms, side-effects of treatment, and the normal process of ageing. Additionally, there is a need for refined index measuring comorbidity in elderly cancer patients. Clearly, more information is needed on the impact of disease and treatment on the quality of life of elderly cancer patients. Such information can contribute significantly to establishing appropriate treatment policy for elderly cancer patients.

Key points

- Few prospective studies have examined optimal psychosocial management of elderly patients with cancer.
- There may be reluctance in the elderly to undergo screening or report bodily changes leading to delay in diagnosis and more advanced stage.
- Reluctance may arise from fear of loss of independence.
- Doctors may mistake symptoms of cancer for those of ageing or not refer cases because of a misconceived notion of protection.
- Quality of life measurements are best sought from the patient and need to be both simple and brief.
- Controlled trials are required to establish the needs of elderly patients undergoing both curative and palliative treatment, together with compliance with therapy.

References

Aaronson, N.K. (1988). Quality of life: what is it? How should it be measured? *Oncology*, **2** (5), 69–74.

Aaronson, N.K. (1991). Methodologic issues in assessing the quality of life of cancer patients. *Cancer*, **67**, 844–50.

Aaronson, N.K. and Beckmann, J. (eds.). (1987). *The quality of life of cancer patients*. Raven Press, New York.

Adelman, R.D., Greene, M.G., and Charon, R. (1987). The physician–elderly patient–companion triad in the medical encounter: the development of a conceptual framework and research agenda. *Gerontologist*, **27**, 729–34.

Allen, K., Cox, E., Manton, K., and Cohen, H.J. (1986). Breast cancer in the elderly: current patterns of care. *Journal of the American Geriatric Society*, **34** (6), 637–42.

American Cancer Society (1990). *Cancer facts and figures—1990*. American Cancer Society, Atlanta, GA.

Bergman, L., Dekker, G., van Kerkhoff, E.H.M., Peterse, H.L., van Dongen,

J.A., and van Leeuwen, F.E. (1991). Influence of age and comorbidity on treatment choice and survival in elderly patients with breast cancer. *Breast Cancer Research Treatment*, **18**, 189–98.

Bergman, L., Dekker, G., van Leeuwen, F.E., Huisman, S.J., van Dam, F.S.A.M. and van Dongen, J.A. (1991). The effect of age on treatment choice and survival in elderly breast cancer patients. *Cancer*, **67**, 2227–34.

Bergman, L., Kluck, H.M., van Leeuwen, F.E., Crommelin, M.A., Dekker, G., Hart, A.A.M., and Coebergh, J.W.W. (1992). The influence of age on treatment choice and survival of elderly breast cancer patients in the southeastern part of The Netherlands: a population based study. *European Journal of Cancer*, **28A**, 1475–80.

Bergman, L., van Dongen, J.A., van Oijen, B., and van Leeuwen, F.E. (1992). Should tamoxifen be a primary treatment choice for elderly breast cancer patients with locoregional disease? (In press.)

Berkanovic, E. and Aaronson, N.K. (1986). The decision to seek medical care for symptoms. *Advances in Health Education and Promotion*, **1**, 165–79.

Berkanovic, E., Hurwicz, M., and Lubben, J.E. (1988). Psychological distress, illness and the decision to seek medical care among the aged. Unpublished final report of grant RO1-AGO5181, funded by the National Institute on Aging, pp. 1–186.

Bonadonna, G. and Valagussa, P. (1981). Dose–response effect of adjuvant chemotherapy in breast cancer. *New England Journal of Medicine*, **304**, 10–21.

Borgers, M.Th.A. and Visser, A.Ph. (1990). Voorlichting aan oudere kanker-patienten. Een literatuuroverzicht. *Tijdschrift voor Sociale Gezondheidszorg*, **68** (3), 137–43.

Cacioppo, J.T., Andersen, B.L., and Turnquist, D.C. (1986). Psychophysiological comparison processes: interpreting cancer symptoms. In *Women with cancer: psychological perspectives*, (ed. B.L. Andersen), pp. 141–67. Springer, New York.

Cacioppo, J.T., Andersen, B.L., Turnquist, D.C., *et al.* (1989). Psychophysiological comparison theory: on the experience, description and assessment of signs and symptoms. *Patient Education and Counseling*, **13**, 257–70.

Cassileth, B.R., Zuplis, R.V., Sutton-Smith, K., and March, V. (1980). Information and participation preferences among cancer patients. *Annals of Internal Medicine*, **92**, 832–6.

Chu, J., Diehr, P., Feigl, L., Glaefke, G., Begg, C., Flicksman, A., and Ford, L. (1987). The effect of age on the care of women with breast cancer in community hospitals. *Journal of Gerontology*, **42**, 185–90.

Crooks, C.E. and Jones, S.D. (1989). Educating women about the importance of breast screenings: the nurse's role. *Cancer Nursing*, **12** (3), 161–4.

Dellafield, M.E. (1988). Informational needs and approaches for early cancer detection in the elderly. *Seminars in Oncology Nursing*, **4** (3), 156–68.

Derdiarian, A. (1987). Informational needs of recently diagnosed cancer patients: part II, method and description. *Cancer Nursing*, **10** (3), 156–63.

Donegan, W.L. (1983). Treatment of breast cancer in the elderly. In *Perspectives on prevention and treatment of cancer in the elderly*, (ed. R. Yancik, P.P. Carbone, W. Bradford Patterson, K. Steel, and W.D. Terry), pp. 83–96. Raven Press, New York.

Feinstein, A.R. and Wells, C.K. (1990). A clinical-severity staging system for patients with lung cancer. *Medicine*, **69**, 1–33.

Feinstein, A.R., Sosin, D.M., and Wells, C.K. (1985). The Will Rogers phenomenon: stage migration and new diagnostic techniques as a source of misleading statistics for survival in cancer. *New England Journal of Medicine*, **312**, 1604–8.

Fentiman, I.S., Tirelli, U., Monfardini, S., Schneider, M., Festen, J., Cognetti, F., and Aapro, M.S. (1990). Cancer in the elderly: why so badly treated? *Lancet*, **335**, 1020–2.

Ferrel, B.A. (1991). Pain management in elderly people. *Journal of the American Geriatric Society*, **39**, 64–73.

Foreman, M.D. and Kleinpell, R. (1990). Assessing the quality of life of elderly persons. *Seminars in Oncology Nursing*, **6** (4), 292–7.

Frank-Stromborg, M. (1988). Future projected trends in the care of the elderly individual with cancer, and implications for nursing. *Seminars in Oncology Nursing*, **4** (3), 224–31.

Ganz, P., Coscarelli Schag, C., and Heinrich, R.L. (1985). The psychosocial impact of cancer on the elderly: a comparison with younger patients. *Journal of the American Geriatric Society*, **33**, 429–35.

Given, B. and Given, C.W. (1989). Cancer nursing for the elderly. A target for research. *Cancer Nursing*, **12** (2), 71–7.

G.I.V.I.O. (Interdisciplinary Group for Cancer Care Evaluation) (1988). Survey of treatment of primary breast cancer in Italy. *British Journal of Cancer*, **57**, 630–4.

Goldsmith, G. and Brodwick, M. (1989). Assessing the functional status of older patients with chronic illness. *Family Medicine*, **21**, 38–41.

Goodwin, J.S., Samet, J.M., Key, C.R., Humble, C., Kutvirt, D., and Hurt, C. (1986). Stage at diagnosis varies with the age of the patient. *Journal of the American Geriatric Society*, **34**, 20–6.

Greene, M.G., Hoffman, S., Charon, R., and Adelman, R. (1987). Psychosocial concerns in the medical encounter: a comparison of the interactions of doctors with their old and young patients. *Gerontologist*, **27** (2), 164–8.

Greenfield, S., Bianco, D.M., Elashoff, R.M., and Ganz, P.A. (1987). Patterns of care related to age of breast cancer patients. *Journal of the American Medical Association*, **257**, 2766–70.

Hahn, D.E.E. and van Dam, F.S.A.M. (1988). Psychosocial implications and complications of cancer treatment in the elderly. In *Cancer in the elderly*, (ed. W.W. ten Bokkel Huinink), pp. 16–31. Excerpta Medica, Elsevier, Amsterdam.

Hahn, D.E.E., de Haan, R., van Dam, F.S.A.M., and Everaerd, W.T.A.M. (1990). Early symptom perception and delay in elderly cancer patients. *International Journal of Cancer Research and Clinical Oncology*; **116**, 56.

Hall, S.W. (1984). Cancer: special considerations in older patients. *Geriatrics*, **39** (7), 74–8.

Haug, M. (1979). Doctor–patient relationship and the older patient. *Journal of Gerontology*, **14** (6), 852–60.

Haynes, R., Taylor, D., and Sackett, D. (1979). *Complicance in health care*. John Hopkins, Baltimore, MD.

Hobbs, P., George, W., and Sellwood, R. (1980). Acceptors and rejectors of an intervention to undergo screening compared with those who referred themselves. *Journal of Epidemiological Community Health*, **34**, 19–22.

Holland, J.C. (1989). Fears and arousal reactions to cancer in physically healthy individuals. In *Handbook of Psychooncology*, (ed. J.C. Holland and R.H. Rowland), pp. 13–21. Oxford University Press.

Holland J.C. and Massie, M.J. (1987). Psychosocial aspects of cancer in the elderly. *Clinics in Geriatric Medicine*, **3** (3), 533–9.

Holmes, F.F. (1983). Aging and cancer. *Recent Results in Cancer Research*, **87**, 1–75.

Holmes, F.F. and Hearne, E. (1981). Cancer stage-to-age relationship: implications for cancer screening in the elderly. *Journal of the American Geriatric Society*, **29**, 55–7.

Host, H. and Lund, E. (1986). Age as a prognostic factor in breast cancer. *Cancer*, **57**, 2217–21.

Kahn, R.L., Antonucci, T.C., and Akiyama, H. (1989). In *Cancer in the elderly*, (ed. R. Yancik and J.W. Yates), pp. 40–5. Springer, New York.

Kane, R.A. (1983). Coordination of cancer treatment and social support for the elderly. In *Perspectives on prevention and treatment of cancer in the elderly*, (ed. R. Yancik, P.P. Carbone, W. Bradford Patterson, K. Steel, and W.D. Terry), pp. 227–38. Raven Press, New York.

Kegeles, S. and Grady, K. (1982). Behavioural dimensions. In *Cancer epidemiology and prevention*, (ed. D. Schottenfeld and J. Fraumeni), pp. 1049–63. Saunders, Philadelphia, PA.

Keller, M.L., Leventhal, H., and Prohaska, T.R. (1989). Beliefs about aging and illness in a community sample. *Research in Nursing and Health*, **12**, 247–55.

Kennedy, J.J. (1991). Needed: clinical trials for older patients. *Journal of Clinical Oncology*, **9** (5), 718–20.

Van Leeuwen, F.E. and Coebergh, J.W.W. (1988). Descriptive epidemiology of cancer in the elderly in the Netherlands. In *Cancer in the elderly*, (ed. W.W. ten Bokkel Huinink), pp. 6–15. Excerpta Medica, Elsevier, Amsterdam.

Leventhal, E.A. (1984). Aging and the perception of illness. *Research in Aging*, **6**, 119–35.

Leventhal, E.A. and Prohaska, T.R. (1986). Age, symptom interpretation, and health behaviour. *Journal of the American Geriatrics Society*, **34**, 185–91.

Leventhal, H., Leventhal, E.A., and Schaefer, P. (1989). Vigilant coping and health behaviour : a life span problem. In *Aging, health and behaviour*, (ed. M. Ory and R. Abeles). John Hopkins, Baltimore, MD.

Levy, S.M. (1983). The aging patient: behavioural research issues. In *Perspectives on prevention and treatment of cancer in the elderly*, (ed. R. Yancik, P.P. Carbone, W.B. Patterson, K. Steel, and W.D. Terry), pp. 237–47. Raven Press, New York.

Love, N. (1991). Why patients delay seeking care for cancer symptoms? *Cancer Evaluation*, **89** (March), 151–7.

McNeil, K.J., Stones, M.J., and Kozma, A. (1986). Subjective well-being in later life : issues concerning measurement and prediction. *Social Indicators Research*, **18**, 35–70.

McNeil, B.J., Pavik, S.G., Sox, H.C., and Tversky, A. (1988). On the elicitation of preferences for alternative therapies. In *Judgement and decision making: an interdisciplinary reader*, (ed. R. Arkes Hal and K.R. Hammond), pp. 386–93. Cambridge University Press.

Mages, N.L. and Mendelsohn, G.A. (1979). Effects of cancer on patients' lives : a personological approach. In *Health psychology*, (ed. G.C. Stone, F. Cohen, and N.E. Adler), pp. 225–84. Jossey-Bass, San Francisco, CA.

Maisiak, R., Gams, R., Le, E., and Jones, B. (1983). In *Progress in clinical and biological research*, (ed. P.F. Engstrom, P.N. Anderson, and L.F. Mortenson), pp. 395–403. Alan R. Riss, New York.

Massie, M.J. and Holland, J.C. (1989). The older patient with cancer. In *Handbook of psychooncology*, (ed. J.C. Holland and J.H. Rowland) pp. 444–54. Oxford University Press.

Menko, F.H. (1986). Voorlichting aan patienten met kanker; ervaringen van hulpverleners. Unpublished doctoral thesis. Free University of Amsterdam.

Mold, J.W., Nevins, M.A., Sherman, F.T., and Waltman, A.C. (1991). Zwakte bij ouderen, waarop wijst dat? *Patient Care*, 47–59.

Mor, V., Masterson-Allen, S., and Goldberg, R.J. (1985). Relationship between age at diagnosis and treatment received by cancer patients. *Journal of the American Geriatric Society*, **33**, 585–9.

Mor, V., Guadagnoli, E., Silliman, R.A., Weitberg, A., Glicksman, A., Goldberg, R., et al. (1989). Influence of old age, performance status, medical and psychosocial status on management of cancer patients. In *Cancer in the elderly: approaches to early detection and treatment*, (ed. R. Yancik and J.W. Yates), pp. 127–46. Springer, New York.

Mueller, C.B., Ames, F., and Anderson, G.D. (1978). Breast cancer in 3,559 women: age as a significant determinant in the rate of dying and causes of death. *Surgery*, **83**, 123–32.

Nerenz, D.R., Love, R.R., Leventhal, H., and Easterling, V. (1986). Psychosocial consequences of cancer chemotherapy for elderly patients. *Health Services Research*, **20** (Suppl. 6), 961–76.

Nylenna, M. (1986). *Cancer—a challenge to the general practitioner*. The Norwegian Cancer Society, Oslo.

Osoba, D. (1991). *The effect of cancer on quality of life*. CRC Press, Toronto.

Ouslander, J.G. and Beck, J.C. (1982). Defining the health problems of the elderly. *Annual Review of Public Health*, **3**, 55–83.

Pearlman, R.A. and Jonsen, A. (1985). The use of quality-of-life considerations in medical decision making. *Journal of the American Geriatrics Society*, **33**, 344–52.

Pearlman, R.A. and Uhlmann, R.F. (1988). Quality of life in chronic diseases: perceptions of elderly patients. *Journal of Gerontology*, **43**, M25–30.

Plagge, I.M.C.M. (1990). Voorlichting aan oudere kankerpatienten. *Medisch Contact*, **45** (40), 1185–7.

Prohaska, T.R., Keller, M.L., Leventhal, E.A., and Leventhal, H. (1987). Impact of symptoms and ageing attribution on emotion and coping. *Health Psychology*, **6** (6), 499–578.

Riley, M.W. (1983). Cancer and the life course. In *Perspectives on prevention and treatment of cancer in the elderly*, (ed. R. Yancik, P.P. Carbone, W.B. Patterson, K. Steel, and W.D. Terry), pp. 25–32. Raven Press, New York.

Rimer, B.K., Jones, W., Wilson, C., Bennett, D.J., and Engstrom, P. (1983). Planning a cancer control program for older citizens. *Gerontologist*, **23**, 384–9.

Rimer, B., Keintz, M.K., and Fleisher, L. (1986). Process and impact of a health communications program. *Health Education Research*, **1**, 29–36.

Safer, M.A., Tharps, Q.L., Jackson, T.C., and Leventhal, H. (1979). Determinants of the stages of delay in seeking care at a medical clinic. *Medical Care*, **17** (1), 11–29.

Samet, J., Hunt, W.C., Key, C., Humble, C.G., and Goodwin, J.J. (1986). Choice of cancer therapy varies with age of patients. *Journal of the American Medical Association*, **255**, 3385–90.

Silliman, R.A., Guadagnoli, E., Weitberg, A.B., and Mor, V. (1989). Age as a predictor diagnostic and initial treatment intensity in newly diagnosed breast cancer patients. *Journal of Gerontology*, **44**, 46–50.

Sprangers, M.A.G. and Aaronson, N.K. (1992). The role of health care providers and significant others in evaluating the quality of life of patients with chronic disease: a review. *Journal of Clinical Epidemiology*, **45** (7), 743–60.

Spilker, B. (1990). *Quality of life assessments in clinical trials*. Raven Press, New York.

Timko, C. (1987). Seeking medical care for a breast cancer symptom: determinants of intentions to engage in prompt or delay behaviour. *Health Psychology*, **6** (4), 305–28.

Tirelli, U., Aapro, M., Obrist, R., Festen, J., Schneider, M., Fentiman, I., and Monfardini, S. (1991). Cancer treatment and old people. *Lancet*, **338**, 114.

Warnecke, R.B. (1989). The elderly as a target group for prevention and early detection of cancer. In *Cancer in the elderly: approaches to early detection and treatment*, (ed. R. Yancik and J.W. Yates), pp. 3–14. Springer, New York.

Weinrich, S.P. and Weinrich, M.C. (1986). Cancer knowledge among elderly individuals. *Cancer Nursing*, **6**, 301–7.

Weinrich, S.P., Blesch, K.S., Dickson, G.W., Nussbaum, J.S., and Watson, E.J. (1989). Timely detection of colorectal cancer in the elderly. *Cancer Nursing*, **12** (3), 170–6.

Williams, T.F. (1990). A perspective on quality of life and care for older people. In *Quality of life assessments in clinical trials*, (ed. B. Spilker), pp. 217–23. Raven Press, New York.

Winograd, H.C. (1986). Geriatric medicine and the cancer patient. *Frontiers of Radiation Therapy and Oncology*, **20**, 27–37.

Wolinsky, F.D., Arnold, C.L., and Nallopati, I.V. (1988). Explaining the decline rate of physician utilization among the oldest-old. *Medical Care*, **26**, 544–53.

Yancik, R. (1983). Frame of reference: old age as the context for the prevention and treatment of cancer. In *Perspectives on prevention and treatment of cancer in the elderly* (ed. R. Yancik, P.P. Carbone, W. Bradford Patterson, K. Steel, and W.D. Terry), pp. 5–19. Raven Press, New York.

Yancik, R. (1989). Introduction: cancer in elderly, an overview. In *Cancer in the elderly: approaches to early detection and treatment*, (ed. R. Yancik and J.W. Yates), pp. xv–xxi. Springer, New York.

Yancik, R. and Ries, L.G. (1989). Caring for elderly cancer patients. Quality assurance considerations. *Cancer*, **64**, 335–41.

Yancik, R., Kessler, L., and Yates, J.W. (1988). The elderly population. Opportunities for cancer prevention and detection. *Cancer*, **62** (8), 1823–8.

Yancik, R., Ries, L.G., and Yates, J.W. (1989). Breast cancer in aging women. A population-based study of contrasts in stage, surgery and survival. *Cancer*, **63**, 976–81.

3

Radiotherapy

P. Scalliet, E. van der Schueren, and D. van den Weyngaert

Introduction

The simple fact that this book and the present chapter have been written proceeds from the belief that old people constitute a distinct group of the population, deserving special consideration in terms of care, diagnostic procedures, and treatments. However, epidemiological data do not define such a separate group; there is rather a continuum of increasing cancer incidence with age, with no clear-cut boundary between young, mature, or aged adults.

From the socio-economic point of view, being old is often associated with retirement and economic non-productivity, both characteristics identifying a particular age class in our society. This distinction has no real counterpart in oncology. Indeed, there is little evidence that a digestive or lung cancer arising at the age of 75 or 80 is any different from the same tumour arising at the age of 50, yet the general belief is that such tumours should be treated differently, more gently, in old people weakened by age and with shorter life expectancy.

The main issues, when dealing with radiation oncology in the elderly, are related to tolerance and to strategies with, as background, the supposedly reduced aggressive behaviour of tumours in older patients and the reduced ability of older patients to support radical irradiation. Surprisingly few data are available in the field of radiation oncology dealing specifically with the indications and tolerance of radiotherapy in the elderly, because either the subject has not been properly investigated up to now or the studies attempting to relate age to tolerance produced negative results which were not considered worth publishing. We suspect the second reason has played a prominent role in the lack of specific knowledge about the age effect in radiotherapy. After all, radiation oncologists have always treated old patients, and data acquired on tissue tolerance are already based on the observation of old patients.

Preliminary remarks

The true effect of age is often obscured by the presence of other associated pathological conditions which are known to influence tolerance to radiotherapy. Of course, aged patients are more exposed to other chronic diseases such as diabetes, hypertension, and heart or lung insufficiency, which may complicate the clinical course of their cancer and impair their ability to sustain a long and sometimes aggressive curative treatment. However, this is only indirectly related to age since impaired vital functions are simply more frequent in aged patients, and are not an intrinsic feature of the elderly. Clearly, the respective effects of age and of chronic diseases other than cancer on treatment tolerance need to be assessed separately.

This has been illustrated, for instance, in a short study on complications following panabdominal radiotherapy. Radiation ileitis developed in 5 out of 39 women irradiated for stage I to III ovarian epithelial tumours (age range 31–78). The most decisive predisposing factors for developing a complication were first an intraperitoneal tumour recurrence (4 out of 5), secondly extensive abdominal surgery prior to radiotherapy (5 out of 5), and finally previous cisplatinum-containing chemotherapy (4 out of 5), although the last factor was of borderline significance only. Age did not play any role since the mean ages of the total group, the subgroup with recurrence, and those with complications were identical (De Winter *et al.* 1990). Another study came to a very similar conclusion (Daly *et al.* 1989).

Treatment strategy in relation to age: radiotherapy as an alternative for patients with medical contraindications for radical surgery

No guidelines are established for adapting treatment strategies to patients' ages (except perhaps in haematology), yet older patients, as already mentioned, are often treated in a different way from younger patients. Indeed, the few available relevant studies do not support the view that less aggressive treatments are indicated in old patients. A recent trial, in which women older than 75 were randomized to receive either tamoxifen alone or surgery, concluded that a mastectomy is the appropriate treatment also for elderly patients with operable breast cancer (Robertson *et al.* 1988). Giving tamoxifen as unique treatment only delayed the moment when surgery became necessary, thus surgery might be carried out under more adverse conditions. In another retrospective series, a preferential allocation of aged women to conservative breast surgery with or without radiotherapy was examined. The latter policy can no longer be advocated since the recurrence rate exceeded 25 per cent compared with only 7 per cent if postoperative radiotherapy was given (Kantorowitz *et al.* 1989).

For cancers in certain locations, curative radiotherapy is preferred to radical surgery, even for tumour stages where, in general, excision would be the treatment of choice. As an example, there is always a small proportion of lung cancer patients with resectable tumours but between 70 and 80 years a patient will not be considered for surgical treatment because of age alone and will be referred to the radiation oncologist. There is evidence that radiotherapy is also curative in some (Nordijk *et al.* 1988), but whether this is an equivalent treatment option remains unknown since comparative data are so far unavailable.

The situation is very similar in bladder cancer. Patients in surgical series are often younger on average and benefit from thorough peroperative staging procedures of which non-surgical series are deprived. An age-based strategy is therefore difficult to establish retrospectively since the quality of staging is not homogeneous among different treatment modalities, with a positive bias in favour of the surgical series (Raghavan *et al.* 1990).

Good treatment begins with good disease staging. However, recent studies have demonstrated that patients aged 75 and older had significantly reduced intensive clinical staging workups (and also less use of surgical or radiation therapies) in prostate, lung, and breast cancer compared with younger patients (Bennett *et al.* 1991; Bergman *et al.* 1991; Greenfield *et al.* 1987; Mor *et al.* 1985). Most of the time, this attitude could not be justified by sound scientific arguments and seems to have been highly influenced by personal opinions of physicians and by the type of hospital in which they practised.

It is sometimes said that less aggressive, gentler treatments are indicated in older age groups since aged patients often present with a more advanced disease. However, this is less than clear and available data are to a large extent inconsistent. In multivariate analysis age is sometimes found to be correlated with more advanced cancer stages (Goodwin *et al.* 1986), and sometimes not (Bennett *et al.* 1991; Bergman *et al.* 1991), and at least one study has found an inverse correlation between age and stage (Samet *et al.* 1990).

Moreover, older patients are more often treated without histological verification of cancer, the diagnosis being simply based on the clinical evidence. Retrospective studies trying to correlate age with treatment outcome are therefore potentially biased. This has been addressed specifically in a paper on patterns of care related to age of men with prostate cancer (Bennett *et al.* 1991). Bone scans were performed significantly more often among the younger patients than in the men aged 65–74 or those older than 75 years. In addition, surgical staging with pelvic lymphadenectomy was carried out significantly more frequently in the youngest patients. Other studies on breast cancer came to similar conclusions (Bergman *et al.* 1991; Greenfield *et al.* 1987).

The fact that older patients may die from intercurrent disease rather than from their tumour does not mean that they do not suffer from the cancer-related morbidity before ultimately dying from other causes. Less aggressive strategies based on the higher incidence of intercurrent disease (Papillon 1990) must then be introduced very carefully, with a thorough appreciation of the potential cancer morbidity which will result from a less intensive therapy. Other strategies, based on the belief that cosmesis is a less prominent concern in older patients (as in Abbatucci *et al.* 1989) are also not supported by data; there is no evidence that the body image is less important with advancing age.

A last important problem, with regard to strategy, is the dilemma concerning palliative versus radical treatment. A precise definition of what is a palliative treatment in this context is a challenging intellectual exercise, but it may tentatively be put in the following way: in some circumstances, a palliative treatment schedule (with lesser burden) will give sufficient tumour regression to cover the expected survival time, a more radical treatment still being theoretically possible. Thus palliation alone may be offered to some patients whose life expectancy is obviously short or very short, but for non-oncological reasons. Looking back at our own practice, there is less than a handful of such patients each year. They all share the same handicap: the association of advanced age and a severe degree of dementia, so that the minimum of active collaboration necessary for the treatment to be delivered in good conditions is not guaranteed. There is also a common feeling with such patients that their quality of life is already impaired to such an extent that attempts to prolong life seem meaningless. However, this attitude is mainly related to philosophical convictions of the doctor, nursing staff, and, it is to be hoped, of the patient themselves, as assessed during conversations with the patients or their family if the patient is demented. This is indeed a very delicate matter, at the shifting boundary between treatment abstention and euthanasia . . .

Is tolerance to radiotherapy age related?

Clinical data are either not mature enough or do not exist as to whether radiation tolerance depends on age, independently of any associated chronic disease. As already discussed, this may either be due to a lack of specific studies or be because attempts to relate age to a different side-effect pattern were inconclusive and hence not published, at least not as separate articles with 'age' as a keyword which would have changed a labour-intensive literature review into an easy task.

We reviewed two years of publications in *Radiotherapy and Oncology*, between January 1989 and December 1990 (24 issues), looking at articles in which any data specifically related to the age effect could be identified. Two hundred and twenty papers were published, of which 107 dealt with the clinical application of radiation, of chemotherapy, or of a combination of both. The other articles were either technical papers (physics) or radiobiological papers (of which none dealt with the age effect).

Of the few randomized studies presented, none used age as a stratification factor. On the contrary, old age was sometimes an exclusion factor (Sarrazin *et al.* 1989). In the other papers, mainly retrospective studies, age was frequently mentioned as a descriptive item of the population investigated, but was not used further in the analysis. Of the 107 clinical papers, 34 were directly concerned with problems of tolerance and presented series where age could have been tested as a separate prognostic factor for developing any form of complication. Unfortunately, less than a handful of authors actually made use of the available information and tested age as a distinct variable in their analyses.

van Limbergen *et al.* (1989) found no influence of age on the cosmetic outcome of conservative breast cancer treatments in a multivariate analysis of a large group of patients. In contrast, a small negative impact of age on cosmesis was suggested from another series, although this did not reach statistical significance and was, at least to some extent, confounded by variations in extent of surgery (Steeves *et al.* 1989). The paper of Daly *et al.* (1989), mentioned above, analysed a series of 188 women irradiated with large pelvic fields who presented later with radiation ileitis. Age had no influence on the probability of developing a complication, but obese women or those older than 75 years were treated systematically with 1.8 Gy instead of 2 Gy per fraction. Another paper, dealing with the risk of developing a brachial plexus injury after radiotherapy for breast cancer found no correlation with age (449 patients with age range 18–92, Powell *et al.* 1990). The only predictor for plexus injury was the use of large daily radiation doses. Thus there was not a single article mentioning age as an independent factor predictive for the development of complications after irradiation.

Seven additional articles discussed treatment strategy or treatment outcome according to age (Abbatucci *et al.* 1989; Bataini *et al.* 1989; Brada *et al.* 1989; Kantorowitz *et al.* 1989; Lydbeert *et al.* 1989; Papillon 1990; Perez *et al.* 1989). Some of these have already been discussed in the previous section.

Despite all this, it is still often claimed that radiotherapy is badly tolerated in the elderly, and that curative irradiations entail more severe acute reactions in old patients, especially in head and neck cancers. For this reason, we reviewed charts of 277 patients with stage I to IV (AJC) head and neck tumours treated with external radiotherapy and salvage surgery between 1979 and 1990. Factors studied included survival, body weight loss during radiotherapy, and possible interruptions of treatment due to acute reactions (van den Weyngaert *et al.* 1991). Body weight loss ensuing from difficulties in food intake was taken as an objective index of acute mucosal reactions to irradiation. The mean age was 63.3 years (range 29–91). For the purpose of the analysis, the patient population was divided arbitrarily into three groups: less than 60 years (40 per cent), 60–70 years (30 per cent), and more than 70 years (30 per cent). Less than 5 per cent of patients needed hospitalization during treatment for artificial nasogastric feeding (independently of age).

The mean body weight loss during radiotherapy was about 4 per cent, but with large variations, some patients gaining weight during their treatment (up to + 8 per cent). Maximal body weight loss was 15 per cent. If any trend was recognizable it was a slight positive effect of age on treatment tolerance. It may seem surprising that patients can gain weight during radiotherapy, but for some patients the contact with the hospital and the medical team was an occasion to improve their overall food intake, both in quantity and in quality, after years of alcohol over-consumption.

No significant influence of age could be found on overall treatment duration or on survival for the different tumour localizations. Thus there was no evidence from our experience that radiotherapy was less well tolerated or less effective in old patients treated for head and neck tumours.

Considering the paucity of relevant data it was interesting to check whether experimental laboratory findings could compensate for the lack of clinical information. Radiobiological data have already been reviewed in a recent paper (Scalliet 1991) and the least that can be said is that they are equally scanty. Most of the laboratory studies explored the age effect by comparing immature with young adult animals. There is indeed a major concern about the tolerance of children to which these investigations were addressed, but comparisons of young with old mature animals could not be found. However, it should be recognized that radiobiological experiments with old animals may entail many biological and logistic problems. Working with old mice or rats which are close to the end of their lifespan is in general quite impractical for all radiation effects and in particular for late effects because they have many other diseases (kidney, spontaneous tumours) and the latency for the observation of late effects then exceeds their life expectancy (the latency in animals and humans is almost the same). Moreover housing animals is very expensive.

Few *in vitro* data deal with the age of cell cultures. Human fibroblasts were found to have similar survival parameters independently of the donor's age (range 11–78 years, Little *et al*. 1988), although the proliferative potential of such cultures clearly decreases with age (Martin *et al*. 1970). In contrast, the rate of DNA damage removal has been found to decrease with increasing age in rat skin (Sargent and Burns 1985), but this observation has no obvious *in vivo* counterpart (Denekamp 1975). However, these projects did not explore skin reactions in very old animals.

Interesting data were published very recently dealing with the effect of host age on microenvironmental heterogeneity of EMT6 tumours implanted into young and ageing mice. The mice may be considered to have been truly old, since they were kept for between 15 and 18 months before starting the experiments. The most striking point is that, although the implanted tumours were identical, the radiobiological hypoxic fraction was much higher in old than in young animals (41 per cent versus 19 per cent), with an obvious impact on the curability. Thus mitomycin C, an agent with selective toxicity to hypoxic cells, produced a greater antineoplastic effect on tumours in ageing mice when used alone or as an adjunct to radiation (Rockwell *et al.* 1991). However, these data contrast with the available clinical evidence that tumour radiocurability at various ages does not seem to differ significantly, as discussed previously.

Logistic problems

The main problem for the elderly patient with radiotherapy is the need for daily trips, five times a week, from home to the hospital. Having only the weekend for resting is often not sufficient because weekends are often dedicated to visits from or to the family and other social activities, equally tiring (if not more so). However, accepting one or more leisure days in the week invariably prolongs the treatment which may be detrimental to its efficacy. Alternatively, larger daily fractions are sometimes advocated in order both to keep the treatment duration within acceptable limits and to offer an additional rest day during the week. This entails a higher risk of late effects, and trading the late effects for a better immediate tolerance requires careful assessment of all factors involved.

Hospitalization is probably of no real help because deprivation of the home surroundings may give rise to disorientation and depression which is even worse. There is no other guideline to advocate than to appreciate on a personal and familial basis which solution is adapted to each particular case, and the consequences of all possible options need to be discussed openly before any treatment decision is made.

Conclusions

Most frequently, tumour related factors (stage, histology, etc.) will override considerations of age in the choice of treatment, if a curative option exists. Many patients who are not eligible for a major operation or for major chemotherapy can tolerate the alternative of radical radiotherapy quite well, and there is no suggestion that age will have a major influence on tolerance. Indeed, a substantial proportion of patients in radiotherapy are over 70 years old, and patients over 80 years old are regularly seen.

'Reduced' treatment is never a solution, unless the life-expectancy of the patient is obviously so poor that the tumour recurrence is unlikely to occur or at least to produce substantial morbidity before the patient has died from another cause.

Key points

- Few data are available on radiation indication and tolerance in the elderly.
- Radiotherapy may be used in the place of radical surgery but the efficacy of this approach has not been studied by randomized trials.
- From limited data, age does not relate to complications after irradiation.
- Attention to fractionation may enable cancer treatment of the elderly but the morbidity of hypofractionation is not yet determined.

References

Abbatucci, J.S., Boulier, N., Laforge, T., and Lozier, J.C. (1989). Radiation therapy of skin carcinomas: results of a hypofractionated irradiation schedule in 675 cases followed for more than 2 years. *Radiotherapy and Oncology*, **14**, 113–20.

Bataini, J.P., Asselain, B., Jaulerry, Ch., Brunin, F., Bernier, J., Ponvert, D., and Lave, C. (1989). A multivariate primary tumour control analysis in 465 patients treated by radical radiotherapy for cancer on the tonsillar region: clinical and treatment parameters as prognostic factors. *Radiotherapy and Oncology*, **14**, 265–78.

Bennett, C.L., Greenfield, S., Aronow, H., Ganz, P., Vogelzang, N.J., and Elashoff, R. (1991). Patterns of care related to age of men with prostate cancer. *Cancer*, **67**, 2633–41.

Bergman, L., Dekker, G., van Leeuwen, F.E., Huisman, S.J., van Dam, F.S.A.M., and van Dongen, A. (1991). The effect of age on treatment choice and survival in elderly breast cancer patients. Cancer, **67**, 2227–34.

Brada, M., Ashley, S., Nicholls, J., Wist, E., Colman, T.J., McElwain, T.J., Selby, P., Peckham, M.J., and Horwich, A. (1989). Stage III Hodgkin's disease – long term results following chemotherapy, radiotherapy and combined modality therapy. *Radiotherapy and Oncology*, **14**, 185–98.

Daly, N.J., Izar, F., Bachaud, J-M., and Delannes, M. (1989). The incidence of severe chronic ileitis after abdominal and/or pelvic external irradiation with high energy photon beams. *Radiotherapy and Oncology*, **14**, 287–95.

Denekamp, J. (1975). Residual radiation damage in mouse skin 5 to 8 months after irradiation. *Radiology*, **115**, 191–5.

De Winter, K., van den Weyngaert, D., Becquart, D., and Scalliet, P. (1990). Panabdominal radiotherapy in ovarian carcinoma: a retrospective analysis of survival and complications (abstract). In *Proceedings of the 9th Annual Meeting ESTRO, Montecattini*, p. 157. ESTRO.

Goodwin, J.S., Samet, J.M., Key, C.R., Humble, C., Kutvirt, D., and Hunt, C.

(1986). Stage at diagnosis of cancer varies with age of the patient. *Journal of the American Geriatric Society*, **34**, 20–6.

Greenfield, S., Blanco, D., Elashoff, R.F., and Ganz, P.A. (1987). Patterns of care related to age of breast cancer patients. *Journal of the American Medical Association*, **257**, 2766–70.

Kantorowitz, D.A., Poulter, C.A., Rubin, P., Patterson, E., Sobel, S.H., Sischy, B., Dvoretsky, P.M., Michalak, W.A., and Doane, K. (1989). Treatment of breast cancer with segmental mastectomy alone or segmental mastectomy plus radiation. *Radiotherapy and Oncology*, **15**, 141–50.

van Limbergen, E., Rijnders, A., van der Schueren, E., Lerut, T., and Christiaens, R. (1989). Cosmetic evaluation of breast conserving treatment for mammary cancer: 2, a quantitative analysis of the influence of radiation dose, fractionation schedules and surgical treatment techniques on cosmetic results. *Radiotherapy and Oncology*, **16**, 253–68.

Little, J.B., Nove, J., Strong, L.C., and Nichols, W.W. (1988). Survival of human diploid skin fibroblasts from normal individuals after X-irradiation. *International Journal of Radiation Biology*, **54**, 899–910.

Lybeert, M.L.M., van Putten, W.L.J., Ribot, J.G., and Crommelin, M.A. (1989). Endometrial carcinoma: high-dose rate brachytherapy in combination with external irradiation; a multivariate analysis of relapse. *Radiotherapy and Oncology*, **16**, 245–52.

Martin, G.H., Sprague, C.A., and Epstein, C.J. (1970). Replicative life span of cultivated human cells. Effects of donors age, tissue and genotype. *Laboratory Investigations*, **23**, 86–92.

Mor, V., Masterson-Allen, S., and Goldberg, R.J. (1985). Relationship between age at diagnosis and treatments received by cancer patients. *Journal of the American Geriatric Society*, **33**, 585–9.

Noordijk, E.M., Vander Poest Clement, E., Hermans, J., Wever, A.M.J., and Leer, J.W.H. (1988). Radiotherapy as an alternative to surgery in elderly patients with resectable lung cancer. *Radiotherapy and Oncology*, **13**, 83–9.

Papillon, J. (1990). Present status of radiation therapy in the conservative management of rectal cancer. *Radiotherapy and Oncology*, **17**, 275–84.

Perez, C.A., Garcia, D., Simpson, J.R., Zivnuska, F., and Lockett, M.A. (1989). Factors influencing outcome of definitive radiotherapy for localised carcinoma of the prostate. *Radiotherapy and Oncology*, **16**, 1–21.

Powell, S., Cooke, J., and Parson, C. (1990). Radiation-induced brachial plexus injury: follow-up of two different fractionation schedules. *Radiotherapy and Oncology*, **18**, 213–20.

Raghavan, D., Shipley, W.U., Garnick, M.B., Russel, P.J., and Richie, J.P. (1990). Biology and management of bladder cancer. *New England Journal of Medicine*, **332**, 1129–38.

Robertson, J.F.R., Todd, J.H., Ellis, L.O., Elston, C.W., and Blamey, R.W. (1988). Comparison of mastectomy with tamoxifen for treating elderly patients with operable breast cancer. *British Medical Journal*, **297**, 510–14.

Rockwell, S., Hughes, C.S., and Kennedy, K.A. (1991). Effect of host age on microenvironmental heterogeneity and efficacy of combined modality therapy in solid tumors. *International Journal of Radiation Oncology, Biology and Physics*, **20**, 259–63.

Samet, J.M., Hunt, W.C., and Goodwin, J.S. (1990). Determinants of cancer stage; a population-based study of elderly New Mexicans. *Cancer*, **66**, 1302–7.

Sargent, E.V. and Burns, F.J. (1985). Repair of radiation-induced DNA damage in rat epidermis as a function of age. *Radiation Research*, **102**, 176–81.

Sarrazin, D., Le, M.G., Arriagada, R., Contesso, G., Fontaine, M., Spielman, M., Rochard, F., Le-Chevalier, T., and Lacour, J. (1989). Ten year results of a randomised trial comparing a conservative treatment to mastectomy in early breast cancer. *Radiotherapy and Oncology*, **14**, 177–84.

Scalliet, P. (1991). Radiotherapy in the elderly. *European Journal of Cancer*, **27**, 3–5.

Steeves, R.A., Phromratanapongse, P., Wolberg, W.H., and Tormey, D.C. (1989). Cosmesis and local control after irradiation in women treated conservatively for breast cancer. *Archives of Surgery*, **124**, 1369–73.

van den Weyngaert, D., Scalliet, P., Verstraete, W., van den Heyning, P., and Peeters, L. (1991). Does age modify tolerance to radiotherapy of head and neck tumours? Proceedings of the Sixth European Conference on Clinical Oncology and Cancer Nursing (ECCO6). *European Journal of Cancer*, **Suppl. 2**, S270.

4

Reducing the toxicity of cancer chemotherapy

Silvio Monfardini, Umberto Tirelli, and Vittorina Zagonel

Introduction

Are elderly patients less tolerant of cancer chemotherapy than their younger counterparts? One of the major factors influencing toxicity in old patients is age-related change in organ sensitivity to administered drugs. In particular, clinical experience suggests an age-dependent decrease in the number of bone marrow stem cells which determines the degree of tolerance of patients to cytotoxic agents in general. The increased vulnerability of the lungs to bleomycin in older patients is also well known. Besides these changes in tolerance of target organs, several physiological alterations[1,2] result in modification of the pharmacokinetic behaviour of drugs, which creates an additional source of toxicity in the elderly:

(1) altered volume of distribution, due to loss of some 30 per cent of cells from the tissues compensated by fat deposition and decreased water content;
(2) decreased plasma protein concentration;
(3) diminished blood supply to the organs;
(4) decreased metabolism and biliary excretion due to liver atrophy;
(5) diminished renal excretion due to loss of about 30 per cent of nephrons;
(6) slower drug absorption, even if it is quantitatively generally adequate;
(7) polypharmacy;

Of all these age-related changes, diminished renal clearance might frequently lead to serious adverse drug effects by significantly prolonging drug elimination, with the consequent increase in drug exposure time.

Understanding how patients could be more liable to the toxic effects of chemotherapy does not provide, however, direct information on the clinical relationship between age and toxicity. Here, the rare retrospective studies[3-5] suffer from the fact that elderly patients were often selected for entry in clinical trials on the basis of normal organ function tests.

Therefore they are not representative of the tolerance to chemotherapy of the whole population of elderly cancer patients. It is also well known that overt cardiac, pulmonary, or renal disease is a commoner specific contraindication to the use of cancer drugs in elderly patients. In fact, a relatively low percentage of elderly patients enter clinical trials and the propensity of physicians to attenuate medical cancer treatment in elderly patients has been shown by a number of studies: increasing age is associated with a decreased likelihood of receiving chemotherapy.[6,7] Psychosocial factors other than tolerance *per se* could explain partially why the administration of adequate chemotherapy seems to correlate inversely with age. Additionally, many elderly patients are subject to polypharmacy because of comorbidity.

Despite limited information on age and toxicity, there is evidence that some categories of drugs can be associated with an increased risk of severe toxicity in elderly patients.[7] In particular, excess haematological toxicity has been shown for several myelosuppressive drugs,[4] and there is a general agreement that tolerance is lower for all myelotoxic drugs in elderly patients. Anthracyclines have been found to be associated with a higher risk of cardiac dysfunction, and bleomycin with increased pulmonary toxicity. Because of the great individual variation in organ function impairment as correlated with 'chronological' age, this parameter alone has a limited value for predicting toxicity in an individual patient. However, it is important when dealing with a population of elderly patients undergoing chemotherapy, since in this group of patients functional problems are more frequent than in a younger population.

Thus new treatment strategies for elderly patients should be designed to minimize the toxicity of treatment without loss of efficacy. Elderly patients who are 'functionally' young can be treated according to protocols commonly adopted for the adult population.

In addition, newer therapeutic approaches (biological response modifiers, differentiation inducers, cytokines, anti-idiotype antibodies), which may be less toxic, should be explored as an alternative to conventional chemotherapy, particularly in the elderly.

Dose reduction

Patients older than 70 years are usually excluded from clinical trials of cancer chemotherapy but this does not mean in all instances that conventional chemotherapy is administered. In many instances age is considered an almost overwhelming contraindication to chemotherapy and non-treatment is more frequent with increasing age, while patients undergoing chemotherapy are often treated with reduced dosages. The suggestion that old patients are less tolerant of chemotherapy than younger patients encourages physicians to reduce dose levels, even though attitudes vary among medical oncologists. The practice of dose reduction may be necessary under certain clinical circumstances but reduced dosages may lead not only to reduced toxicity but also to lower response rates.[8] In general, the choice of drug dosage level for elderly patients is between the following:

(1) non-scheduled (empirical) dose reduction;
(2) scheduled dose reduction (a priori reduction);
(3) no dose reduction, initial dose as for younger patients.

In empirical dose reduction a lower and therefore safer starting dose is selected, then escalated subsequently depending on the toxicity induced by preceding courses. The practice of non-scheduled or non-defined dose reduction may be useful for an individual patient, but causes difficulties in the evaluation of treatment activity and toxicity in a patient population treated with the same combination of drugs. Such dose reduction is more an indication of the clinical philosophy of the medical oncologist looking after the patient. Empirical dose attenuation is practised for the following reasons:

(1) There is a general opinion that tolerance to myelosuppressive drugs and potentially cardiotoxic drugs is lower in the elderly.
(2) There is almost general agreement that neurotoxicity is higher.
(3) Drugs excreted renally such as cisplatinum, carboplatin, bleomycin, and methotrexate may carry a heavier risk of unexpected toxicity. The same applies to hepatic excretion and to drugs inactivated by the liver such as doxorubicin, vincristine, and vinblastine.[2]

Scheduled dose reduction for elderly patients has been adopted in the adjuvant chemotherapy setting of breast cancer, as well as in advanced disease.[9] This policy leads to the entry in clinical trials of elderly patients who would otherwise have been excluded. Tolerance to treatment is probably increased, but for breast cancer patients a survival advantage was demonstrated only in those receiving full-dose chemotherapy.[9] Dose reduction based on creatinine clearance of patients with advanced breast cancer treated with cyclophosphamide, methotrexate, and fluorouracil (CMF) has been proposed because of the primary renal excretion of methotrexate and 5-fluorouracil[10]. The toxicity in elderly patients given modified CMF was lower than the toxicity in younger patients given standard CMF, but the response rate of the older patients was reduced.

The possibility that a segment of the population may receive reduced and consequently less effective drug dosages has led to the third policy, that age alone should not be a criterion for dose reduction. This attitude implies that an elderly patient with adequate organ function and without medical complications is no less a candidate for full-dose chemotherapy treatment than younger counterparts. Although this strategy has been adopted by the National Cancer Institute (NCI) and Eastern Cooperative Oncology Group (ECOG) there have been few publications following this policy.[3-5,11] The analysis of severe toxicity in patients over 70 compared with patients under 70 years, performed by Begg and coworkers[3,11] using data from ECOG studies is important since elderly patients were shown to have similar toxicity rates to their younger counterparts, with the exception of methyl carboxy chloroethyl nitrosourea (methyl CCNU) and methotrexate, which were more toxic. However no information was available then as to whether or not the patients over 70 received the same percentage of the drug planned as patients under 70. In these published studies it is clear that there was some selection of elderly cases, since the prevalence of elderly patients in those trials is lower than expected from the general population.

The coexistence of these opposed approaches (non-scheduled and scheduled dose reduction, no dose reduction) derives from the difficulty in predicting toxicity. Approaches to this derive from pharmacokinetics, the assumption being that tolerance to chemotherapy depends primarily on drug exposure. New, relatively simple methods have been devised in plasma concentration/time plots for estimating the area under the curve (AUC) based on limited numbers of plasma samples.[12,13] If a given AUC were associated with a specific degree of toxicity, the plan for elderly patients could be based on determination of the AUC during an initial course of the chosen combination of therapy. Individual patient's AUCs could then be compared with standard values associated with established therapeutic activity and acceptable toxicity. Doses could be adjusted to reach this range in subsequent courses. Pharmacokinetic monitoring could then give a rational approach to dose adjustment in the elderly. This application merits further study.

Protection from toxicity

In the last decade there has been a widespread tendency towards the experimental use of high-dose chemotherapy with or without bone marrow transplantation in an attempt to increase the percentage of complete remissions, and survival, for adult and paediatric patients with a variety of tumour types. Dose escalation and a decreased interval between successive courses have also been tested on adult patients, age being one of the major selection criteria. In elderly patients a reasonable first objective concerning the intensity of treatment would be to deliver a full dose at the usual intervals. Even though there is great variability between individual elderly patients in terms of tolerance to chemotherapy, (and chronological age does not necessarily account for such heterogeneity), patients over the age of 65–70 continue to be excluded from clinical trials because of the perception that they are more prone to severe toxicity. The availability of tools to decrease treatment toxicity would favour the entry of the elderly into controlled clinical trials, resulting in more data being acquired on the treatment tolerance of such patients, and more information being obtained on investigative agents for organ protection.

Bone marrow toxicity

Haemopoietic growth factors (HGFs) represent the most promising means to ameliorate bone marrow damage following cytotoxic chemotherapy. Granulocyte macrophage colony stimulating factor (GM-CSF) and granulocyte colony stimulating factor (G-CSF) decrease the incidence or febrile episodes and reduce the duration of severe neutropenia, and are being used in adult patients for the acceleration of neutrophil recovery following chemotherapy. Recombinant human erythropoietin may be used to correct anaemia in selected cancer patients.[14] However, the question arises of whether the age-related changes in bone marrow function, such as reduction in the number of stem cells and microenvironmental damage, may allow positive action of exogenously administered HGFs. Until recently only sparse and anecdotal experience on the use of GM-CSF and G-CSF in the elderly was available. A fresh review of the English literature comprising 24 studies has shown that, in 67 patients older than 65 years compared with 137 younger patients, age did not prevent a normal response to HGFs.[15] Prospective studies are needed but at least these initial elements suggest that HGFs may actually improve the tolerance for older patients to chemotherapy.

The side-effects described for GM-CSF include fever, arthralgia, bone pain, fatigue, rash, possible fluid retention, and atrial fibrillation. The side-effects of G-CSF are definitely fewer and consist essentially of bone pain. In view of possible cardiac complications it is obvious that in elderly patients testing of G-CSF should be preferred to GM-CSF. Erythropoietin is practically devoid of side-effects and its administration is extremely well tolerated.[14]

The current improvements in the understanding of the tertiary molecular structure of haemopoietic growth factors might lead to the engineering of recombinant molecules in which the active sites (such as receptor binding or receptor stimulating domains) are preserved whilst those related to major toxic effects could be deleted.[16,17] Moreover, the modern recombinant DNA technology has allowed the construction of 'hybrid' growth factors (i.e. fusion protein containing active regions of two different haemopoietins) which might show an enhanced biological activity when compared with the individual cytokines.[18,19] A GM-CSF/IL-3 (Interleukin-3) fusion protein has been synthesized recently which contains the active domains of the two molecules. This hybrid protein can induce a significant multilineage haematological stimulation both *in vivo* and *in vitro* in non-human primates.[19] Furthermore, it has been shown that the infusion of such a fusion protein provides long-lasting haemopoietic stimulation, circumventing the need of continuous exposure to the factor which represents a major drawback in the clinical use of individual haemopoietins.

Clinical protection from bone marrow toxicity has been attempted with ethiophos (WR-2721), a compound developed to prevent the side-effects from nuclear explosions. This agent has been employed in ovarian carcinoma[20] with some preliminary evidence of a decrease in the complications deriving from serious haematological toxicity and in advanced head and neck carcinoma.[21] However, WR-2721 produces a mild peripheral neuropathy. No data are yet available for elderly patients treated with this compound.

Cardiac toxicity

The prevention of anthracycline-related cardiac toxicity is a field of ongoing research. Coenzyme Q10,[22] 1-carnitine,[23] ICRF-187 (Imperial Cancer Research Fund)[24,25] and other agents have been tested *in vitro*, but at the moment there is no clinically useful agent available for this purpose. In view of the increasing incidence of age-related cardiac disease, the development of an effective myocardial protection from anthracycline toxicity could be extremely valuable in the elderly.

Renal protection

Age-related renal toxicity has been demonstrated for methotrexate[3] but not for cis-platinum.[26] For this reason glutathione, which reduces renal toxicity following cis-platinum administration,[27] would be predicted to be of limited value in elderly patients. Nevertheless, glutathione allows cis-platinum administration with a minimal hydration, which can be useful in patients who also have left ventricular failure.

Lung toxicity

Drug-related pulmonary fibrosis is also related to pre-existing environmental conditions (air pollution, workplace, hygiene) or to life-style factors (e.g. smoking habits). These risk factors are more prevalent in the elderly, owing to their longer exposure to environmental factors.

Only recently have the mechanisms underlying drug-related pulmonary fibrosis in subjects receiving cytotoxic chemotherapy (bleomycin, cyclophosphamide, nitrosoureas) been explored. Studies have revealed that several cytokines, including transforming growth factor β1 (TGF-β1), tumour necrosis factor (TNF), and IL-1, are involved in drug-stimulated intra-alveolar collagen deposition.[28] Strategies aimed at inhibiting the activity of such cytokines can therefore be designed.

Tailor-made chemotherapy

Specifically devised chemotherapy in the elderly must aim to achieve a significant improvement in the quality of life during the residual life-span rather than merely inducing prolonged survival.

Compared with single agents, the benefits of combination chemotherapy are counterbalanced by an expanded spectrum of toxic effects. For some solid tumours at an advanced stage, such as non-small cell lung cancer, gastrointestinal tract carcinomas, melanomas, head and neck carcinomas, there is no definite evidence indicating the superiority of combination chemotherapy in terms of significant responses and consequent palliation. Therefore, if chemotherapy has to be started in elderly patients affected by the above-mentioned neoplasms, in our opinion it is better to use single agents. Even phase II studies using agents with a novel mechanism of action seem more rational than combination chemotherapy or refinement of known combination chemotherapy regimens.

The approach of single-agent chemotherapy in elderly patients could also be adopted for chemosensitive haematologic neoplasm such as non-Hodgkin's lymphomas (NHL) of favourable histology and multiple myeloma (MM). In another chemosensitive neoplasms of elderly patients, small-cell lung cancer, single-agent chemotherapy has provided superimposable results to those following combination chemotherapy, in the hands of Bork *et al.*[29] These authors found both vepeside (VP-16) and teniposide (VM-26) to be well tolerated and highly active. In addition, Carney *et al.*[30] confirmed the high activity and acceptable toxicity of VP-16 singularly administered to the elderly with advanced small-cell lung cancers.

Since the indications for single-agent chemotherapy in elderly patients are broader than in adults, properly and well-conducted new drug development programmes should consider the elderly specifically. These cases may enable a better evaluation of new drugs than those of adult patients showing progression after combination chemotherapy.

The first combination chemotherapy regimens specifically designed for elderly patients were used for non-Hodgkin's lymphoma and are described in detail in Chapter 10. These regimens include anthracyclines presumed to be less toxic than doxorubicin (pirarubicin, epidoxorubicin) or mitoxantrone, while VP-16 and VM-26 are substituted for vincristine. In gastric carcinoma Wilke *et al.* have been using VP-16 in association with 5-fluorouracil plus Leucovorin instead of doxorubicin and mitomycin C to avoid expected cardiac toxicity.[31] If combination chemotherapy regimens are to be used for advanced breast carcinoma no longer responsive to endocrine therapy, mitoxantrone or epiadriamycin should probably be substituted for doxorubicin.

To reduce chemotherapy-related toxicity, changes in drug scheduling need to be considered. For example, it has been shown that a continuous intravenous infusion reduces toxicity significantly to target organs for drugs such as anthracyclines, vincristine, and bleomycin. However, such an approach may result in reduced compliance because of the need for prolonged hospitalization.

A specific effort must also be devoted to the development of therapeutic regimens which can be given on an out-patient basis, such as the use of orally administered cytotoxic drugs. The advantage provided by such regimens derives from a reduced social cost, better compliance in elderly patients, and improved control of side-effects and toxicity.

Future directions

Along with the development of chemotherapy specifically devised for elderly patients, special emphasis must be devoted to the testing in this age group of new therapeutic approaches exploiting recent advances in the understanding of tumour cell biochemistry and immunology, and in the recombinant DNA technology.

Differentiation-inducing agents

The therapeutic use of so-called 'differentiation-inducing agents' has the aim of causing the terminal maturation of clonogenic cells, leading to the clonal extinction of the tumour.[32]

This approach has mainly been exploited for the treatment of acute myeloid leukaemias (AML) or myelodysplastic syndromes (MDS) of the elderly. Following the conflicting results obtained with low doses of cytosine arabinoside (ARA-C),[33] recent studies have established that other differentiation inducers including all-trans retinoic acid (ATRA)[34] and the DNA-methyltransferase inhibitors 5-azacytidine (5-azaCR)[35] and 5-aza 2'-deoxycytidine (5-azaCdR, decitabine)[36,37] are able to induce clinical remission in leukaemic patients with a reduced toxicity. Different investigators have shown that ATRA induces complete haematological remissions in patients with acute promyelocytic leukaemia, including those who relapsed after previous chemotherapy, with minimal toxicity and also reduces the onset of disseminated intravascular coagulation in these subjects.[34]

Two important studies from Cancer and Leukemia Group B (CALGB)[35] and from an Italian study group[37] have indicated that both 5-azaCR and decitabine show promising clinical activity with acceptable toxicity in the elderly with MDS and AML. Both studies demonstrated that these DNA-methyltransferase inhibitors can induce differentiation of tumour cells and restore bone marrow functions in MDS patients, displaying an activity similar, if not superior, to that of HGFs.[35] Differentiation therapy has the advantage of very low cost compared with recombinant molecules, and may also be administered on an out-patient basis (oral or subcutaneous route).

Cytokines

The clinical use of recombinant cytokines or other biological response modifiers has the aim of causing direct destruction of tumour cells or their indirect elimination by the host immune system.

Recombinant α-2 interferon (α-IFN) has shown definite activity in the treatment of MM or low grade NHL, allowing a reduction in chemotherapy courses required to achieve remission, and consequently a reduction in drug-related adverse effects. A significant increase in disease-free survival was also observed in MM.[39] α-IFN activity in solid tumours has not been clearly established and studies of combination therapy with cytotoxic drugs are ongoing. Other cytokines able to induce tumour cell destruction or immune modulation such as IL-2 at a low dosage and M-CSF are under study and may also offer the advantage of an effective therapy and acceptable toxicity.[16] As an example, IL-2 might be used at a very low dosage in conjunction with monoclonal anti-idiotypic antibodies to obtain a maximal antitumour effect and avoid the life-threatening complications linked to its use at a high dosage. However, while M-CSF can activate the function of immune cells it has also shown a cytostatic action on melanoma and endometrial cancer cells.[16]

Anti-idiotype antibodies

Anti-idiotypic monoclonal antibodies are able to elicit an antitumour effect via the triggering of an 'idiotypic cascade' (active specific immunotherapy) or to induce direct tumour inhibition (passive specific immunotherapy) with minimal or no toxicity.[40]

For patients with follicular lymphomas, anti-idiotypic monoclonal antibodies against surface immunoglobulines have been employed successfully in treatment.[41] Significant tumour shrinkage with minimal toxicity was achieved in most of the patients, including those treated previously with chemotherapy. Even though the emergence of idiotypic negative tumour cell variants (present before antibody therapy) might hamper the curative impact of such an approach, it provides a unique non-toxic tool to induce substantial tumour shrinkage and consequently to reduce the cytotoxic drug load necessary to achieve eradication of the disease. In this way a further reduction in drug-related adverse effects may be obtained. The advantage of such an approach in the elderly is evident. Alternatively, anti-idiotypic antibodies might induce a significant antitumour immunoresponse (both humoral and cellular), activating an 'idiotypic network' so bypassing tumour-induced tolerance or immune suppression.[40]

Studies in patients with colon cancer and melanoma[42] have clearly demonstrated that it is possible to trigger the 'idiotypic cascade'.

Biochemical modulation

The biochemical modulation of the cellular metabolism of anticancer drugs might offer a useful tool to reduce organ toxicity without losing therapeutic effectiveness.

ARA-C, for example, is an effective agent in the treatment of AML. A major limitation to the cytotoxic efficacy is its rapid deamination to uracil arabinoside (ARA-U). High dosage ARA-C (HDA) has been used successfully to circumvent ARA-C deamination and to improve therapeutic efficacy.[43] Although HDA has a definite beneficial effect in AML patients who are resistant or refractory to conventional doses of ARA-C, it is associated with severe extrahaematological toxicity, mainly neurotoxicity. HDA-induced neurotoxicity is related to the accumulation of high ARA-U levels in the central nervous system (CNS) and is particularly severe in elderly patients who display a prolonged CNS clearance of ARA-U.[44] The use of non-toxic inhibitors of the cytidine deaminase enzyme may represent a strategy to reduce toxicity and to exploit the whole potential clinical efficacy of ARA-C in elderly patients with AML. The reduction of ARA-U formation might in fact result in reduced neurotoxicity and prolonged exposure of tumour cells to ARA-C. Clinical use of one such inhibitor, tetrahydrouridine (THU), has been recently proposed in cancer patients who were treated with conventional dose ARA-C+THU coinfusion.[45] This combination therapy resulted in plasma ARA-C levels comparable with those achieved using HDA (greater than 10 μM) with a concomitant reduction of ARA-U levels and absence of neurotoxicity.[45] These results, if confirmed in a large series of patients, might represent a clear example of biochemical modulation of drug efficacy/toxicity to be applied in the elderly with ARA-C responsive tumours.

Cell resistance

The use of agents able to modify cancer cell resistance to chemotherapy may represent an additional tool to reduce organ damage from cytotoxic drugs. The discovery of the link between multidrug resistance (MDR) and the different isoforms of membrane p-glycoproteins, encoded for by the MDR family of genes in humans, has opened a new avenue in the modulation of cancer chemotherapy.[46] It is conceivable that the clinical use of agents able to circumvent MDR would not only allow complete eradication of the tumour cells, but might also result in a diminished need for high-dose chemotherapy (HDCT). HDCT is in fact one of the currently adopted strategies to eliminate resistant tumour cell clones. Although potentially effective in several instances, HDCT is usually associated with life-threatening toxicity.

Clinical trials of resistance-modifying agents (RMAS) have now started and include the combined use of chemotherapy plus RMAS such as verapamil, cyclosporine A, and phenothiazines.[46] Preliminary results of such studies have indicated that there was no dramatic increase in toxicity for normal tissues expressing high levels of p-glycoprotein such as renal, hepatic, and intestine epithelium. In addition, there was no evidence that RMAS potentiated the acute toxicity of the cytotoxic drugs. Promising clinical results have been obtained in NHL and in AML with the use of cyclosporin A,[47] and in MM patients treated with verapamil plus vincristine, adriamycin, dexametuasone (VAD) chemotherapy even though in the latter study verapamil-related cardiac alterations have been reported in some of the patients.

Conclusions

Elderly patients continue to be excluded from clinical trials because of the perception that they are prone to develop more severe toxicity to chemotherapeutic agents. Some authors have stressed that great variability exists between individual elderly patients in terms of tolerance to treatment, pointing out that chronological age does not necessarily account for such heterogeneity.[3,11] Others contend that taking the whole population of elderly patients potentially amenable with cancer chemotherapy (instead of a selected segment of subjects with optimal organ function), tolerance is reduced mainly because of the accompanying age-related pathology. In this evolving scenario the development of methods to decrease toxicity in elderly patients requires their entry into prospective clinical trials. This will result immediately in better clinical practice with a consequent increase in the collection of accurate data on treatment tolerance of the elderly.

Increased tolerance to cancer chemotherapy in the elderly can be achieved by selecting those patients who can possibly tolerate the same regimes used for young adults, by reducing drug dosages in those with decreased organ function, by improving bone marrow depression by administration of haemopoietic growth factors, and by designing regimens which can decrease the risk of cardiac, neurological, and renal toxicity. Possibly the use of cytokines, differentiating agents, and anti-idiotypic antibodies, with or without concomitant attenuated cytotoxic chemotherapy, will increase the fraction of elderly patients receiving adequate anticancer therapy over the next few years.

In our opinion the era of the 'enlightened empiricism' in anticancer drug prescription should eventually come to a conclusion, since there are several promising approaches allowing the design of prospective trials for elderly patients or at least leading to more uniform treatment of such patients.

Key points

- Physiological changes of ageing may affect the pharmacokinetics of anticancer drugs.
- Pharmacokinetic studies are needed in the elderly to relate drug concentration to toxicity and achieve scheduled rather than non-scheduled reductions in dosage in specific patients.

Acknowledgement

The authors would like to acknowledge the support of the Italian Association for Cancer Research (AIRC) and the Research Project Applicazioni Cliniche della Ricerca Oncologica of the Italian National Centre for Research (CNR).

References

1. Zagonel, V., Carbone, A., Farpel-Forins, S., Kuse, R., Pelic, S., Monfardini, S., *et al.* (1990). Management of non-Hodgkin's lymphoma in elderly patients. In *Management of non-Hodgkin's lymphomas in Europe*, (ed. S. Monfardini), pp. 35–44. Springer, Berlin.
2. Kelly, J.F. (1986). Clinical pharmacology of chemotherapeutic agents in old age. *Frontiers of Radiation Therapy and Oncology*, **20**, 101–11. In *Cancer and the elderly*, (ed. J.M. Vaeth and J. Meyer). Karger, Basel.
3. Begg, C.B. and Carbone, P.P. (1983). Clinical trials and drug toxicity in the elderly. The experience of the Eastern Cooperative Oncology Group. *Cancer*, **52**, 1986–92.
4. Begg, C.B., Elson, P.J., and Carbone, P.P. (1989). A study of excess hematologic toxicity in elderly patients treated on cancer chemotherapy protocols. In *Cancer in the elderly*, (ed. R. Yancik and J.W. Yates), pp. 149–63. Springer, New York.
5. Leslie, W., Bonomi, P., Gale, M., Thomas, C., Taylor, S.G. IV, Reddy, S., *et al.* (1991). Combined modality therapy for Stage III non-small cell lung cancer:

comparison of geriatric and younger patients. *Proceedings of the American Society of Clinical Oncology*, **10** (904), 261.

6. Mor, V., Masterson-Allen, S., Goldberg, R.J., Cummings, F.J., Glicksman, A.J., and Fretwell M.D. (1985). Relationship between age at diagnosis and treatment received by cancer patients. *Journal of the American Geriatrics Society*, **33**, 585–9.

7. Walsh, S.J., Begg, C.B., and Carbone, P.P. (1989). Cancer chemotherapy in the elderly. *Seminars in Oncology*, **16** (1), 66–75.

8. Frei, E. and Canellos, G.P. (1980). Dose: a critical factor in cancer chemotherapy. *American Journal of Medicine*, **69**, 585–94.

9. Bonadonna, G. and Valagussa, P. (1981). Dose response effects of adjuvant chemotherapy in breast cancer. *New England Journal of Medicine*, **304**, 10–15.

10. Gelman, R. and Taylor, S.G. (1984). Cyclophosphamide, methotrexate and 5-fluorouracil chemotherapy in women more than 65 years old with advanced breast cancer: the elimination of age trends in toxicity by using doses based on creatinine clearance. *Journal of Clinical Oncology*, **2**, 1404–13.

11. Begg, C.B., Cohen, J.L., and Ellerton, J. (1980). Are the elderly predisposed to toxicity from cancer chemotherapy? An investigation using data from the East Cooperative Oncology Group. *Cancer Clinical Trials*, **3**, 369–74.

12. Collins, J.M., Grieshaber, C.K., and Chabner, B.A. (1990). Pharmacologically-guided phase I trials based upon preclinical development. *Journal of the National Cancer Institute*, **82**, 1321–26.

13. Ratain, J.M., Schilsky, R.L., Choi, K.E., Guarnieri, C., Grimmer, D., Vogelzang, N.T., Senerjian, E., and Liebne, M.A. (1989). Adaptive control of etoposide administration: impact of interpatient pharmacodynamic variability. *Clinical Pharmacology and Therapeutics*, **45** (3), 226–33.

14. Abels, R., Gordon, D., Rose, E., Carey, R., Yazujian, D., and Rudnick, S. (1991). Efficacy and safety of recombinant human erythropoietin in anaemic cancer patients. *Proceedings of the American Society of Clinical Oncology*, **10** (1230), 346.

15. Shank, W. and Balducci, L. (1991). Recombinant hemopoietic growth factors may protect older patients from chemotherapy myelodepression. *Proceedings of the American Society of Clinical Oncology*, **10** (1150), 326.

16. Mertelsmann, R. and Herrman, F. (eds.) (1990). *Hematopoietic growth factors in clinical applications*. Dekker, New York.

17. Mertelsmann, R. (1991). Hematopoietins: biology, pathophysiology and potential as therapeutic agents. *Annals of Oncology*, **2**, 251–63.

18. Moore, M.A.S. (1991). The future of a cytokine combination therapy. *Cancer*, **67**, 2718–26.

19. Williams, D.E. and Park, L.S. (1991). Hematopoietic effects of a granulocyte-macrophage colony—stimulating factor/interleukin-3 fusion protein. *Cancer*, **67**, 2705–7.

20. Kemp, G.M., Glover, D.J., and Schein, P.S. (1990). The role of WR-2721 in the reduction of combined cis-platin and cyclophosphamide toxicity. *Proceedings of the American Society of Clinical Oncology*, **9** (259), 67.

21. Kish, J.A., Ensley, J.F., Tapazoglou, E., and Al-Sarraf, M. (1990). Evaluation of the chemoprotective effect of WR-2721 in recurrent and advanced head and neck cancer patients. *Proceedings of the American Society of Clinical Oncology*, **9** (697), 180.

22. Mortensen, S.A., Aabo, K., Jonson, T., and Baandrup, U. (1986). Clinical and non-invasive assessment of anthracycline cardiotoxicity: perspectives on myocardial protection. *International Journal of Clinical Pharmacology Research*, **6** (2), 137–50.
23. Neri, B., Compani, T., Multani, A., and Torcia, M. (1983). Protective effects of 1-carnitine on acute adriamycin and daunomycin cardiotoxicity in cancer patients. *Clinical Trials Journal*, **20** (2), 98.
24. Herman, E., Ardalan, B., Bier, C., Waravdekar, V., and Krop, S. (1979). Reduction of daunorubicin lethality and myocardial cellular alterations by pretreatment with ICRF-187 in Syrian golden hamsters. *Cancer Treatment Reports*, **63**, 89–92.
25. Shipp, N.G., Dorr, R.T., Alberts, D.S., and Liebler, D. (1991). Mitoxantrone cardiotoxicity in vitro is selectively blocked by ICRF-187. *Proceedings of the American Society of Clinical Oncology*, **10** (340), 118.
26. Hrushesky, W.J.M., Shimp, W., and Kennedy, B.J. (1984). Lack of age-dependent cis-platin nephrotoxicity. *American Journal of Medicine*, **76**, 579–84.
27. Di Re, F., Bohm, S., Oriana, S., Spatti, G.B., and Zunino, F. (1990). Efficacy and safety of high-dose cisplatin and cyclophosphamide with glutathrone protection in the treatment of bulky advanced epithelial cancer. *Cancer Chemotherapy Pharmacology*, **25**, 355–60.
28. Lazo, J.S. and Dale, G.H. (1990). The molecular basis of interstitial pulmonary fibrosis caused by antineoplastic agents. *Cancer Treatment Reviews*, **17**, 165–7.
29. Bork, E., Hansen, M., Ersbol, P., Dombernowsky, P., Bergman, B., and Hansen, H.H. (1989). A randomised study of teniposide (VM-26) versus vepeside (VP-16) as single agents in previously untreated patients with small cell lung cancer > 70 years. *Proceedings of the American Society of Clinical Oncology*, **8** (890) 229.
30. Carney, D.N., Grogen, L., Smit, F., Harford, P., Berendsen, H.H., and Postmus, P.E. (1990). Single-agent oral etoposide for elderly small cell lung cancer patients. *Seminars in Oncology*, **17** (suppl. 2), 44–53.
31. Wilke, H., Preusser, P., Fink, U., Achterrath, W., Meyer, H.J., Stahl, M., Lenaz, L., Meyer, J., Siewert, J.R., and Geerlings, H. (1990). New developments in the treatment of gastric carcinoma. *Seminars in Oncology*, **17** (suppl. 2), 61–70.
32. Lotan, R., Francis, G.E., Freeman, C.S., and Waxman, S. (1990). Differentiation therapy. *Cancer Research*, **50**, 3453–64.
33. Cheson, B.D., Jasperse, D.M., Simon, R., and Friedman, M.A. (1986). A critical appraisal of low dose cytosine arabinoside in patients with acute non-lymphocytic leukemia and myelodysplastic syndromes. *Journal of Clinical Oncology*, **4**, 1857–64.
34. Castaigne, S., Chomienne, C., Daniel, M.T., Ballerini, P., Berger, R., and Fenaux, P. (1990). All-trans retinoic acid as a differentiation therapy for acute promyelocytic leukaemia: I, clinical results. *Blood*, **76** (9), 1704–09.
35. Silverman, L.R., Holland, J.F., Nelson, D., Clamon, G., Powell, B.L., Bloomfield, C.D., *et al.* (1991). Trilineage (TRL) response of myelodysplastic syndromes (MDS) to subcutaneous (SQ) azacytidine. *Proceedings of the American Society of Clinical Oncology*, **10** (747), 222.
36. Pinto, A., Attadia, V., Fusco, A., Spada, O.A., and Di Fiore, P. (1984).

5-aza-2'-deoxycytidine induces terminal differentiation of leukemic blasts from patients with acute myeloid leukemia. *Blood*, **64**, 922–9.

37. Pinto, A., Zagonel, V., Attadia, V., Bullian, P.L., Gattei, V., Carbone, A., Monfardini, S., and Colombatti, A. (1989). 5-aza-2'-deoxycytidine as a differentiation inducer in acute myeloid leukaemias and myelodysplastic syndromes of the elderly. *Bone Marrow Transplantation*, **4** (3), 28–32.

38. Zagonel, V., Lore, G., Marotta, G., Babare, R., Sardeo, G., Gattei, V. *et al.* (1993). 5-Aza-2¹-Deoxycytidine (Decitabine) induces trilineage response in unfavourable myelodysplastic syndromes. *Leukemia*, **7**, Suppl. Monograph 1, 30–5.

39. Mandelli, F., Avvisati, G., Amadori, S., Boccadoro, M., Gernone, A., Lauta, V.M., Marmont, F., Petrucci, M.T., Tribacio, M., Vegna, M.L., Dammacco, F., and Pileri, A. (1990). Maintenance treatment with recombinant interferon alfa-2b in patients with multiple myeloma responding to conventional induction chemotherapy. *New England Journal of Medicine*, **322**, 1430–4.

40. Wettendorff, M., Iliopoulos, D., Tempero, M., Kay, D., DeFreitas, E., Koprowski, H., and Herlyn, D. (1989). Idiotypic cascades in cancer patients treated with monoclonal antibody CO17-1A. *Proceedings of the National Academy of Sciences of the United States of America*, **86**, 3787–91.

41. Brown, S.L., Miller, R.A., Horning, S.J., Czerwinski, D., Hart, S.M., McElderry, R., Basham T., Warnice, R.A., Merigan, T.C., and Levy, R. (1989). Treatment of B-cell lymphomas with anti-idiotype antibodies alone in combination with alpha interferon. *Blood*, **73** (3), 651–61.

42. Mittelman, A., Chen, Z.J., Kageshita, T., Yang, H., Yamada, M., Baskind, P., Goldberg, N., Puccio, C., Ahmed, Arlin, Z., and Ferrone, S. (1990). Active specific immunotherapy in patients with melanoma. *Journal of Clinical Investigation*, **86**, 2136–44.

43. Rudnick, S.A., Cadman, E.C., Capizzi, R.L., Skeel, R.T., Bertino, J., and McIntosh, S. (1979). High dose cytosine arabinoside in refractory acute leukemia. *Cancer*, **44**, 1189–93.

44. Lopez, J.A. and Agarwal, R.P. (1984). Acute cerebellar toxicity after high dose cytarabine associated with CNS accumulation of its metabolite, uracil arabinoside. *Cancer Treatment Reports*, **68**, 1309–10.

45. Kreis, W., Chan, K., Budman, D.R., Schulman, P., Allen, S., Woiselberg, L., Lichiman, S., Henderson, V., Freeman, J., Deere, M., Andreeff, M., and Vincieve, R.R.A.V. (1988). Effect of tetrahydrouridine on the clinical pharmacology of 1-B-D-arabinofuranosylcytosine when both drugs are coinfused over three hours. *Cancer Research*, **48**, 1337–42.

46. Nooter, K. and Herweijer H. (1991). Multidrug resistance (MDR) genes in human cancer. *British Journal of Cancer*, **63**, 663–9.

47. Sonneveld, P. and Nooter, K. (1990). Reversal of drug-resistance by cyclosporin-A in a patient with acute myelocytic leukaemia. *British Journal of Haematology*, **75**, 208.

5

Chemotherapy for patients aged over 80

M. Schneider, A. Thyss, P. Ayela, M.H. Gaspard,
J. Otto, and A. Creisson

The suggestion that chemotherapy be offered to patients over 80 years of age may seem unrealistic or even provocative. However, this situation will undoubtedly be encountered more and more frequently as a result of the demographic evolution of the population in Europe. As an example, in 1982 only 2 per cent of men and 5 per cent of women in France were older than 80 years. However, in the Alpes-Maritimes department of France, where Nice is situated, these figures were 4 per cent and 7 per cent respectively. Retirement to warmer regions may lead to a concentration of the elderly which will be reflected in the recruitment of aged patients at particular cancer treatment centres. Between 1986 and 1989, when other French cancer centres were compared with the Antoine Lacassagne Center in Nice, the percentages of new cancers diagnosed in patients over 80 years (excluding skin cancer) were 5 per cent versus 10 per cent for men and 7.5 per cent versus 10 per cent for women.

This group of older patients represents a significant and treatable population. Some will have potentially chemosensitive tumours, and be in otherwise good health without significant comorbidity.

Despite this, the elderly are less likely than younger cancer patients to receive specific antitumour therapies (Walsh *et al.* 1989). The proportion of cancer patients receiving potentially curative therapies declines with chronological age and the frequency of non-treatment increases with advancing age (Samet *et al.* 1986). The percentage of cancer treatment regimens that omit chemotherapy is higher for older than for younger patients (Chu *et al.* 1987). This is especially true for patients over 80 years of age (Exton-Smith *et al.* 1982).

Toxic side-effects are among the treatment-related problems faced by cancer patients receiving chemotherapy (Begg *et al.* 1989). Controlled trials have shown that certain drugs induce more frequent or more severe complications in older patients. However, in most of these trials, age-stratified toxicity data are not reported (Fentiman *et al.* 1990).

The relationship between age and prospects for response and survival in chemotherapy recipients is hard to determine even for a given form of cancer. To date, the results of clinical trials including both elderly and non-elderly patients have not yielded definite conclusions. Moreover, in those clinical trials which included only elderly patients, the comparative performances of age groups within those studies have not been indicated directly. The development of therapeutic strategies for elderly cancer patients is assuming an increasing importance and thus any data derived from treatment of the elderly are useful in formulating management schemes and designing new clinical trials. For this reason, the results summarized in this chapter of chemotherapy treatment of a large series of patients aged over 80 are of interest, in terms of both toxicity and efficacy.

A breakdown of the primary tumour sites of patients aged 80 years or more, seen at the Antoine Lacassagne Cancer Center in Nice, between 1985 and 1987 is given in Table 5.1. In addition it shows the proportion of these patients among the total group of patients treated during this time. In absolute terms, breast cancer was the commonest tumour and those aged over 80 comprised 9 per cent of the total population of breast cancer cases. Colorectal cancers were diagnosed in 36 elderly patients and these comprised 17 per cent of the total of individuals with this primary tumour. Interestingly, 22 patients had unknown primary tumours and these comprised 16 per cent of this group, probably reflecting a reluctance to 'over-investigate' elderly individuals.

Of these 256 new patients aged over 80, chemotherapy was given to 101 (39 per cent), none of whom had previously received cytotoxic chemotherapy. More men than women were treated in this way (54 men and 47 women); the median age was 82 years (range 80–7). Table 5.2 indicates the malignancies treated, and shows that head and neck tumours were common

Table 5.1 New cancer patients aged over 80 at the Nice Cancer Center 1985–7

Type of cancer	Number	Percentage of all patients
Head and neck	38	7.5
Colon–rectum	36	17
Breast	76	9.3
Lung	26	9.8
Bladder	16	6.2
Uterus	26	7.8
Leukaemia–lymphoma	16	8.8
Unknown primary	22	16

Table 5.2 Chemotherapy after 80 years of age: type of cancer

Type of cancer	Men	Women
Head and neck	16	1
Breast	–	14
Bladder	11	5
Colon–rectum	3	9
Lymphoma–leukaemia	13	12
Miscellaneous	11	6
Total	54	47

+ 11 different types of cancer.

among men and breast cancer was common among women. Lymphomas and leukaemias treated by chemotherapy were equally common among both sexes.

Chemotherapy protocols

The most commonly used chemotherapy agents are summarized in Table 5.3 which shows that 5-fluorouracil (5FU) was administered to more than half of the patients. Polychemotherapy was more frequent than monochemotherapy (72 per cent versus 28 per cent). Four or more drugs were given to 15 patients (15 per cent).

Table 5.3 Chemotherapy after 80 years of age: most commonly used drugs and combinations

Drug or number of agents	Number	%
5FU	53	53
Cis-platinum	36	36
Anthracyclines	24	24
Vinca alkaloids	17	17
Cyclophosphamide	17	17
1 agent	28	28
2 agents	39	39
3 agents	19	19
4 agents or more	15	15

Table 5.4 Chemotherapy after 80 years of age: percentage of theoretical doses actually administered to 79 patients with multi-agent chemotherapy

Percentage of theoretical dose	Number of patients	% %	Dose reduction at the 2nd or 3rd cycle
100–90	23	29	3
89–80	8	10	1
79–70	16	20	4
69–60	23	29	1
59–50	7	9	–
<50	2	9	–

A variety of chemotherapy protocols was in use, depending on the tumour type. Patients with non-Hodgkin's lymphoma were given either COP (cyclophosphamide 750 mg/m^2 day 1, vincristine 1.4 mg/m^2 day 1 and prednisone 40 mg/m^2 days 1–5, every 3 weeks), CHOP (COP + doxorubicin 50 mg/m^2 day 1), or LAC (aclacynomycin 30 mg/m^2 days 1–4, etoposide 60 mg/m^2 day 2, cyclophosphamide 500 mg/m^2 day 4 and prednisone 40 mg/m^2 days 1–4, every 3 weeks).

For patients with acute myeloid leukaemia either low dose Ara-C (10 mg/m^2 every 12 hours days 1–20), or a combination of Ara-C (120 mg/m^2 days 1–7) and doxorubicin (30 mg/m^2 days 1–3), was used. Anthracyclines were used for the treatment of metastatic breast cancer, either doxorubicin (12 mg/m^2 per week) or mitoxantrone (8 mg/m^2 every 3 weeks). A combination of cis-platinum (80 mg/m^2 day 1) and 5FU (800 mg/m^2 days 2–6) was given as a continuous infusion every three weeks for patients with head and neck and bladder cancers.

All dosages were 20 per cent lower than those used in adults of average age. For those individuals aged over 85 years doses were reduced by a further 10 per cent.

Results of treatment

The median number of courses administered was three with a range of 1–32. One course of treatment was given to 32 patients (32 per cent), and two courses to 14 (14 per cent). Three courses were given to 21 (21 per cent) and four or more to 34 (34 per cent). Thus over half the patients (55 per cent) received three or more courses.

Sixty per cent of the patients received more than 70 per cent of the theoretical dose administered in adults of average age. Dose reductions, in particular during the first cycle, were often empirical. Dose reductions during the second and third cycles were infrequent, and were not correlated with the doses administered during the first cycle (Table 5.4).

Only moderate haematological, gastrointestinal, and renal toxicities were noted in the total of 380 chemotherapy cycles. The 13 per cent haematological toxicity rate included 5.5 per cent grade 3 or 4 toxicity. There were only six cases of grade 3 mucositis and/or vomiting, and one case of altered renal function (Table 5.5). In addition to four unexplained sudden deaths, two deaths were attributable to the toxicity of the chemotherapy. Severe and rapid alteration of general status occurred in three patients.

Finally, among patients with head and neck and bladder cancers, myocardial ischaemia occurred in three of the 91 cycles associating cis-platinum and 5FU (that is, in 3 per cent of these cycles). By comparison, the frequency of such ischaemia in average age adults is only around 2 per cent and, globally, toxicity was very similar in patients over 80 years of age compared with the overall population of patients. A systemic pharmacokinetic study at the half-way point in the 5FU cycle prompted reduction of the 5FU dose in only 10 per cent of the cycles (area under the curve greater than 15 000 ng/ml per hour).

Response to treatment was observed in 19 patients. Table 5.6 shows the correlations between performance status and response to chemotherapy. The majority of these responders (16/19) had a performance status of 0 or 1. If survival is analysed as a function of response to chemotherapy, median survival was 5.5 months in non-responders, 13.5 months in stabilized patients, and 15.5 months in the responders.

In the light of these results, it appears worthwhile obtaining disease stabilization with chemotherapy. For example, over half of the patients with metastatic breast cancer (7/13) survived more than 1 year (12+, 14, 15, 15+, 21, 36, 67+ months). Survival was also correlated with performance status. The response rate of our patients with non-Hodgkin's lymphomas

Table 5.5 Chemotherapy after 80 years of age: toxicity, 380 cycles

	WHO grade				Total
	1	2	3	4	
Haematological	7	21	14	7	49 (13%)
Nausea–vomiting mucositis	1	6	6	–	13 (3.4%)
Renal	1	2	1	–	4 (1%)

Table 5.6 Chemotherapy after 80 years of age: correlation between response and performance status (WHO)

Performance status	Complete response	Partial response	Failure or non-evaluable	Total
0	2	2	11	15
1	5	7	36	48
2	2	1	21	24
3	–	–	14	14
Total	9	10	82	101

was 20 per cent and one-third of these patients lived longer than 1 year (Table 5.7). Two partial responses were achieved with the MOPP regimen (Methotrexate Vincristine, Procarbazine, Prednisone), these two patients with stage IIB and IVB disease had fairly acceptable survivals of 10 and 34 months respectively. With three cycles of cis-platinum 5FU, a response was obtained in 31 per cent of the 16 patients aged over 80 with head and neck cancers (one complete and four partial responses). By comparison,

Table 5.7 Malignant lymphomas after 80 years of age: non-Hodgkin's lymphoma

Treatment	Percentage of dose	Number of cycles	Myelotoxicity	Response	Survival (months)
1 COP	100	1	2	NR	30
2 COP	100	1	3	NE	57
3 COP	100	1	NE	NE	9
4 COP	60	2	0	NR	14
5 COP	100	6	2	RC	15
6 COP	50	1	NE	NE	3
7 COP	80	3	0	NR	16
8 COP	100	3	2	PR	6
9 COP	100	6	0	NR	6+
10 CHOP	60	2	4	NR	7
11 CHOP	100	3	4	NE	2
12 CHOP	100	1	NE	NE	4
13 LAC	100	6	0	PR	7
14 LAC	100	5	2	NE	10
15 LAC	100	6	0	NR	6

COP, cyclophosphamide, vincristine and prednisone; CHOP, cyclophosphaimde, adriamycin, vincristine and prednisone; LAC, aclacynomycin.

response rates were 79 per cent in 54 patients aged between 70 and 79 years (28 complete responses and 15 partial responses), and 87 per cent in the entire study population (105 complete responses and 90 partial responses out of 225).

This type of study indicates that advanced chronological age alone is not sufficient justification for decisions to limit or withhold treatment, including chemotherapy. In place of chronological age, it is suggested that the clinician consider a range of patient characteristics when formulating optimal plans of cancer therapy: physiological age, performance status, level of tumor aggressivity, presence or absence of intercurrent illness, and quality of life. More research is needed to establish firmly the limitations of chronological age as a diagnostic and prognostic tool in the treatment of cancer.

Finally, chemotherapy is feasible in patients aged over 80 years of age. Responses can be obtained with prolongation of survival in responders and stabilized patients. There are important variations in tolerance but chemotherapy is dangerous in patients with bad performance status. Protocols must be adapted for each case and tolerance and efficacy must be evaluated after each cycle. Clinical trials in elderly patients with age stratification are needed.

Key points

- There are major variations in tolerance and poor performance status is a poor contraindication.
- Worthwhile prolongation of survival and stabilization of disease can be achieved in very elderly patients given chemotherapy.

References

Begg, C.B., Elson, P.J., and Carbone, P.P. (1989). A study of excess hematologic toxicity in elderly patients treated on cancer chemotherapy protocol. In *Cancer in the elderly: approaches to early detection and treatment*, (ed. R. Yancik). Springer, Berlin.

Carbone, A., Volpe, R., Gloghini, A., Trovo, M., Zagonel, V., Tirelli, U., and Monfardini, S. (1990). Non-Hodgkin's lymphoma in the elderly: pathologic features at presentation. *Cancer*, **66**, 1991–4.

Chu, J., Diehr, P., Feigl, P., Glaefice, G., Begg, C., Glicksman, A., and Ford, L. (1987). The effect of age on the case of women with breast cancer in community hospital. *Journal of Gerontology*, **42**, 185–90.

Exton-Smith, A.N. (1982). Epidemiological studies in the elderly: methodological considerations. *American Journal of Clinical Nutrition*, **35**, 1273–9.

Fentiman, I.S., Tirelli, U., Monfardini, S., Schneider, M., Festen, J., Cognetti,

F., and Aapro, M.S. (1990). Cancer in the elderly: why so badly treated? *Lancet*, **I**, 1020.

Samet, J., Hunt, W.C., Key, C., Humble, C.E., and Goodwin, J.S. (1986). Choice of cancer therapy varies with age of patient. *Journal of the American Medical Association*, **255**, 3385–90.

Walsh, S.J., Begg, C.B., and Carbone, P.P. (1989). Cancer chemotherapy in the elderly. *Seminars in Oncology*, **16**, 66–75.

6

Breast cancer: principles of management

A.P.M. Forrest and I.S. Fentiman

Incidence

Breast cancer remains the most common form of neoplasia affecting western women and their most frequent cause of death from malignant disease. In the west, risk of the disease is further increased by age, pointing to an environmental cause. However, young women are not exempt. After the age of 25 years, its incidence rises steeply in the premenopausal years. During the menopause there is a transient downward 'hook', following which the incidence continues to rise, albeit at a slower rate, to reach a peak of 2.5 per 1000 women per annum. As there are fewer older women in the population, the peak prevalence of the disease occurs at age 55; yet 40 per cent of all breast cancers occur in women aged over 70 years.[1]

There is concern that the incidence of breast cancer is gradually increasing. Initially this increase affected younger women, but recently there has been an apparent sharp increase in older women. In a recent study from the USA, in which the overall increase in incidence between the time periods 1974–8 and 1986–7 was 31 per cent, that in women of 65–74 and 75–84 years of age, during the most recent 5 years of data, has been 6 and 9 per cent per annum.[2] Before this increasing incidence can be attributed to a change in the natural history of the disease, it is important to eliminate an apparent increase in incidence due to such factors as increasingly efficient cancer registration, the more frequent use of screening mammography, and changing thresholds for diagnosis. In the study referred to, specific attention was paid to the increasing use of mammography. While this largely accounted for the increase in incidence noted in middle-aged women it did not explain that in the young or elderly groups. This is not surprising; mammography has low sensitivity in young women with dense breasts; while for the elderly, it may not be acceptable as a screening method, even when a personal invitation is sent. If the natural history of the disease in older women is changing, at a time in life when oestrogen levels are at their lowest since puberty, the role of associations such as the increasing use of hormone replacement therapy must be kept under review.

During most of this century breast cancer mortality has generally paralleled incidence, indicating that changes in management, and the controversies which accompanied them, have had little effect on the natural history of the disease. However, there is now some evidence that mortality from the disease may be falling (for example in Sweden and Holland), most probably owing to the impact of mammographic screening and the increasing use of systemic therapy.[3,4] Our concepts of the biology of breast cancer have changed.

Biology of breast cancer

During most of the twentieth century, breast cancer was believed to be primarily a loco-regional disease, confined for much of its time-span within the 'catchment area' of the lymphatics draining the breast and related chest wall. Only when the defensive barrier presented by the regional lymph nodes was overcome did dissemination lead to the development of 'secondary' deposits of tumour in bones and viscera, a phenomenon believed to be preventable by radical loco-regional therapy, even at a stage when the lymph nodes were known to be involved by metastatic disease. Such 'curative' therapy involved removal of the breast, the underlying muscles of the chest wall, and the axillary lymph nodes, sometimes extended to include removal of the internal mammary or supraclavicular lymph nodes and often supplemented by radical radiation of these other regional nodal areas. However, 'cure' was then defined as freedom from 'recurrence' of the disease within 5 years.

With reports of longer periods of follow-up after such treatment, the falsehood of this premise became apparent, and it was appreciated that in the majority of women symptomatic invasive breast cancer could not be cured by local treatment alone. Even 30 years following primary treatment, such women continued to die at a rate greater than the normal female population of the same age, this most commonly from metastatic breast cancer.[5] It was apparent that the majority of patients with symptomatic invasive breast cancer already had disseminated disease at the time of primary treatment, although because of the slow proliferative rates of these micrometastases it might be many years before they became clinically evident. Survival could only be improved either by applying local ablative treatment before the disease had disseminated, or by additional systemic therapy to ablate the micrometastatic disease. Both approaches are now known to be effective. Controlled randomized trials of population screening by mammography (to detect breast cancer during its preclinical 'sojourn' phase), and of the use of ovarian ablation, the anti-oestrogen tamoxifen, and chemotherapy as adjuvants to primary local treatment, have shown a significant reduction in mortality from this disease.[6,7]

The elderly

Older women, in general, are less 'bothered' about their health than those with family responsibilities. They may wish to avoid troublesome forms of treatment with their potential morbidity, and they dislike hospitalization, or disruption of their pattern of life by the need for frequent attendances at treatment centres. These attitudes must be taken into account when planning the management of those who develop breast cancer. This can be considered in four stages.

Diagnosis

The key investigations for the diagnosis of breast cancer are clinical examination, mammography, and fine needle aspiration cytology. In older women the diagnosis is usually obvious on clinical examination. A cancer is usually easily palpable within their atrophic breasts and the classical signs of integration into surrounding breast tissue, from infiltration by malignant cells, are readily elicited. Because of delay in presentation the skin may be infiltrated or ulcerated.

A mammogram is also easier to interpret in the elderly woman, the dense spiculated opacity of the cancer standing out clearly against the lucent background of the atrophic fatty breast. Because of the discrete nature of the mass, fine needle aspiration will produce good sampling for cytology, allowing unequivocal confirmation of clinical and mammographic findings.

Benign causes of a breast mass are both less common than in younger women, and are more easily differentiated. A fibroadenoma presents as a smooth hard mass which is well defined from surrounding breast tissue. As it will have been present since youth, central necrosis will have led to the accumulation of gross globular calcifications, visible on the mammogram. Cysts, although rare, can be diagnosed quickly by fine needle aspiration.

Staging

The second phase in the investigation of a patient with breast cancer is to determine the extent of the disease. Local extent is assessed by clinical examination of both breasts and the axillary and supraclavicular regions and should include calliper measurements of the tumour. Mammography of both breasts is necessary to exclude multifocal disease. Methods to detect occult micrometastatic disease, such as bone scans, have poor sensitivity and there is little point in submitting the elderly asymptomatic patient to the discomfort of needless investigation. These can be limited to a careful clinical examination, X-ray of the chest and pelvis, and routine haematological and biochemical tests.

Treatment

As most women presenting with breast cancer at an old age have clinically obvious invasive cancer, consideration must be given to both local and systemic methods of treatment. The supposition that breast cancer in the elderly has a more favourable prognosis than in younger women is not supported by studies of survival following primary local treatment, some of which show the reverse. From the long-term follow-up of 3558 women with breast cancer, 94 per cent treated by local surgery alone, Mueller *et al.* reported that deaths from breast cancer were not only more frequent in the 779 women over the age of 70 years but also occurred more rapidly than in the younger age groups, even when only those with tumours limited to the breast were considered.[8] This finding was supported by a study of 57 068 patients reported from Sweden.[9] Breast cancer is no less aggressive in old age.

Nor is the impression that the disease is diagnosed at a later stage in the elderly borne out by the facts. In a recent review it was shown that 75 per cent of those aged over 75–80 years present with TNM stage I or stage II disease.[10] A summary of data from four studies is given in Table 6.1.[11–14] In addition, as shown in Table 6.2, the percentage of patients with histologically involved axillary nodes is 47 per cent which is similar to that in symptomatic younger women.[13, 16–20]

Table 6.1 TNM stage at presentation in the elderly

Reference	Age	TNM stage			
		I	II	III	IV
11	>80	23 (31)	15 (20)	18 (24)	19 (25)
12	>80	73 (52)	25 (18)	8 (6)	35 (24)
13	>80	39 (25)	77 (49)	24 (15)	16 (10)
14	>75	2907 (53)	1221 (22)	686 (12)	698 (13)
		3042 (52)	1338 (23)	736 (12)	768 (13)

Figures in brackets are percentages.

Table 6.2 Axillary nodal involvement in the elderly with operable breast cancer

Reference	Age	Nodal involvement	
15	>70	133/242	(55)
16	>75	29/44	(66)
17	>75	33/58	(57)
18	>70	22/56	(39)
19	>75	566/1299	(44)
20	>75	85/163	(52)
13	>80	54/116	(47)
		965/2070	(47)

Figures in brackets are percentages.

Local treatment

As indicated above, the orthodox treatment of symptomatic invasive breast cancer should include local and systemic therapy: local treatment to control the disease in the breast and axillary lymph nodes and to prevent local relapse, systemic treatment to destroy or suppress the growth of micrometastases. Mastectomy is still the preferred local treatment by many, this being combined with either surgical clearance of the axilla or with a lower axillary node sample when, should nodal involvement be proven, radiotherapy is normally given. While there are arguments in favour of a sampling procedure in younger women, there is little point in submitting the older woman to both surgery and radiotherapy. If mastectomy is the preferred option a full axillary dissection is preferable. Many surgeons in the USA now suggest that a partial dissection of the axilla (up to the level of the upper border of the pectoralis minor muscle) is sufficient to prevent relapse.

Modifications of total removal of the breast, in which peripheral areas of breast tissue are left behind, are being practised by some surgeons. In our view, if mastectomy is the operation to be performed it should be complete. It must be appreciated that the wish for reconstruction of the breast is not age dependent. While elderly women should not normally be considered for rigorous procedures to reconstruct the breast by myocutaneous flaps, there are a number who appreciate the reconstruction of a breast mound with the sub-pectoral insertion of a sialastic implant. The availability of this simple method of restoring confidence should be brought to their attention.

The alternative method of local therapy now proven, at least for small tumours, to give equal control to mastectomy, is local excision of the tumour, supplemented post-operatively by radical radiation, thus allowing conservation of normal breast. A general anaesthetic is still necessary for the surgical procedure (which includes an axillary lymph node clearance or sample) and as radiotherapy is given in small and repeated fractions, attendance over a period of several weeks is necessary. As it is not certain that a cancer is suitable for conservation treatment until it has been completely examined histopathologically, a second operative procedure may be required. For this reason some elderly women may elect to have the 'one-off' form of treatment offered by mastectomy.

Systemic treatment

Proven methods for the systemic treatment of micrometastases include multimodal cytotoxic chemotherapy (conventionally with cyclophospham-ide, methotrexate, and 5-fluorouracil), and the anti-oestrogen tamoxifen which is suppressive of growth. The overview analysis reported by the Early Breast Cancer Trials Collaborative Group (1992) suggested that chemotherapy gave a better survival advantage in women under 50 years of age, while tamoxifen gave equal benefit in older women.[6] As tamoxifen has few side-effects, most surgeons in the UK prefer to use this agent as an adjuvant to local therapy in women over 50 years of age, although it would now appear that its combination with chemotherapy may give additional benefit. Information on the effect of systemic therapy in elderly women is sparse: all but a few controlled randomized trials exclude women over 70 years of age.

Reduction in extent of therapy

Some surgeons, believing that breast cancer in the elderly runs a relatively benign course, suggest that orthodox treatment by total mastectomy or local excision with radical radiotherapy is unnecessarily radical, and that local excision of the tumour alone without post-operative radiotherapy may be sufficient. The results of two recent studies suggest that this is not so. Relapse rates following local excision of the tumour alone approximated 30 per cent at 4 years.[21,22] The study from Birmingham included 96 women aged 70 years or more with stages I and II cancer, 35 per cent of whom relapsed within 31 (3–120) months. Although most relapses were controlled adequately by tamoxifen therapy, so that further surgery or radiotherapy was seldom required, few would regard a relapse rate of this magnitude as being acceptable. It should be noted that tamoxifen was not given as part of the primary treatment.

Table 6.3 Uncontrolled studies of tamoxifen treatment for elderly patients with operable breast cancer

Reference	Number of patients	Immediate failure		Late failure	
		PD	SD	PR	CR
23	27	5 (19)	2 (7)	2/5	0/15
24	160	27 (17)	38 (2)	7/49	21/46
25	100	10 (10)	22 (22)	2/29	10/39
26	113	24 (21)	34 (3)	37/55	
		(17)	(24)	(20)	

PD progressive disease, SD static disease, PR partial remission, CR complete remission.
Numbers in brackets are percentages.

Primary tamoxifen therapy

Elderly patients may not wish to embark on surgery with general anaesthesia if there is an alternative, simpler treatment. There has been recent interest in giving tamoxifen as primary systemic treatment for operable disease, reserving surgery for those tumours which are uncontrolled or which later relapse. A series of reports has been published during the past decade, which indicates that more than 50 per cent of tumours will regress, half completely, and a further 20 per cent remain static when tamoxifen is given (Table 6.3).[15,23–27] Long-term follow-up indicates that it is those with a complete remission who have the longest duration of response. In the study of 113 women from Dundee, in which the minimum follow-up was 5 years, 87 per cent of the 38 women with a complete remission status had local control of disease at that time.[26] As the side-effects of tamoxifen are minimal, many elderly women, given the option, prefer an initial trial with this drug.

Prediction of response

The time taken to assess the response of the disease to tamoxifen therapy is variable, and in slow-growing disease may take many months. There is natural concern that such delay may introduce additional hazards for the patient should surgical treatment later be required. A prospective study in Edinburgh attempted to determine the relationship between an early assessment of response and eventual worthwhile benefit.[28] To assess early response, the tumour size was measured at frequent intervals by calliper during the first 12 weeks of therapy, and the regression line drawn through the mean diameters was used to determine whether the tumour had increased or decreased in size, or whether it had remained static. However, a reduction in size during this 3 months period predicted only two-thirds of those patients who went on to have worthwhile remission of disease at 2 years. Conversely, an increase in tumour size during the first 3 months of therapy, while denoting failure of control, was observed only in a proportion of those classified as eventual failures. Frequent and prolonged follow-up is therefore a necessity for women treated primarily with tamoxifen therapy alone.

The concentration of oestrogen receptor protein in a tumour is a known predictor of response, but only in a negative sense. Tumours which are poorly endowed with receptor protein have little chance of response to anti-oestrogen therapy. The standard dextran-charcoal saturation analysis (DCC) has proved of little value to predict the response of elderly patients to tamoxifen therapy. In part this is because with ageing, receptor levels in tumours rise exponentially while the incidence of receptor-poor tumours falls.[29] In the Edinburgh report of primary tamoxifen treatment, 37 of the 100 elderly patients had oestrogen receptors estimated by the DCC assay, and only two had receptor-poor (less than 10 fmol/mg protein) tumours. The median ER value in the 35 ER positive tumours was 300 fmol/mg. [25] Additionally, the performance of a biochemical assay requires tumour tissue which, in turn, demands an open biopsy which in patients who elect for primary tamoxifen treatment is counterproductive.

Better discrimination is afforded by the use of the ERICA immunocyto-chemical assay which can be carried out on cells obtained by fine needle aspiration.[30] In a retrospective study of 52 patients, objective remission of disease was described only in one of 21 patients whose tumour contained less than 25 per cent of cells which stained positively for receptor; this result is in agreement with data reported by Coombes *et al.*[31] In our recent prospective study, ERICA assays were available in 46 patients. Only one of 13 who had less than 20 per cent of cells stained had worthwhile benefit from tamoxifen, as defined by control of disease at 2 years.[25] Since in these two studies non-response to tamoxifen therapy was predicted with certainty in 34 out of 98 patients, with false predictions of non-response in only two patients, it is reasonable to suggest that elderly patients with operable disease should be considered for an initial trial of tamoxifen therapy only if an ERICA immunocytochemical assay on a fine needle aspirate demonstrates that over 20 per cent of cells contain oestrogen receptor protein.

Screening

In 1989 the Clinical Practice Committee of the American Geriatrics Society published a position statement on screening for breast cancer in elderly women.[32] There were six recommendations.

1. Women of all ages, including those over the age of 65, should be instructed and encouraged to perform monthly self-examination of the breast.
2. Primary care providers should perform annual breast examination on their elderly female patients.
3. All women over the age of 65 should have a mammogram performed at least every 2 or 3 years until at least the age of 85.
4. The primary care provider who orders the mammogram should be assured that up to date equipment which allows for the lowest radiation dose is used.
5. Departure from the above recommendations is reasonable if the physician together with patient and family takes into account individual circumstances such as life expectancy and medical or physical condition.
6. More research is needed on breast cancer screening in elderly women.

Of all these statements only the final one is inarguably correct. Can a good case be made for screening women aged 65 years or more? At present in Britain, mammographic screening is offered to women aged 50–64 on a triennial basis and those aged 65 years or more may attend, but do not receive invitations to do so.

In a study of a general practice in Manchester, UK, 631 women were identified as being aged 65–79.[33] They were sent an appointment for screening, a prepaid reply card, a leaflet, and a letter of support from their general practitioner. In the same practice there were 738 women aged 50–64 who had been invited to take part in the National Screening Project. Of the younger group 491 (77 per cent) attended as compared with 344 (61 per cent) of those aged 65–79. The pick-up rate for cancers was 4/1000 in the younger group and 11/1000 in the older group in which no negative biopsies were performed. It was suggested that if the results could be repeated nationwide the economics of extending routine call and re-call screening to elderly women should be examined.

Harris *et al.*[34] interviewed 1163 women and 159 physicians in two North Carolina counties to determine their attitudes toward mammographic screening. Of the women in their 30s, 25 per cent had ever had a mammogram and 15 per cent had been X-rayed within the previous year. Of the women in their 50s, 52 per cent had ever had mammograms, and 35 per cent had this within the previous year. Of the women in their 70s, 37 per cent had had a mammogram and 26 per cent had breast x-rays within the previous year. Of the youngest group of women, two-thirds considered themselves to be at increased risk of breast cancer compared with only 30 per cent of those aged over 70.

The clinicians were gynaecologists, family practitioners, and general physicians and 19 per cent ordered mammograms for the majority of their patients aged 30–40, 40 per cent of those aged 40–50, and 14 per cent of those aged 50–74. It was concluded that younger women were being subjected to unnecessary mammograms whereas older women were not being screened because of a misunderstanding of risk by both patients and their doctors. In another study of 576 primary care physicians on the roster of the Indiana State Medical Association similar results were reported.[35] For women without a family history of breast cancer, 63 per cent referred for screening those in their 50s, 70 per cent referred those aged 65, but only 29 per cent referred women of 70. The reasons given for not referring older women included breast self-examination being sufficient, patients refusing mammograms, increased comorbidity, physicians' forgetfulness, and confusion regarding guidelines for mammography.

To evaluate the impact of screening on mortality in various age groups Mandelblatt *et al.* constructed a decision analysis model.[36] Data derived from the Surveillance Epidemiology and End Results (SEER) program for age and stage distribution were used together with screening results from the Breast Cancer Demonstration Detection Project (BCDDP), and women aged 65–74 who took part in the Swedish Two Counties study. Five age categories were examined, 65–69, 70–74, 75–79, 80–84, and over 85. For every age group the model predicted survival benefits. For women aged 65–69, screening extended life by an average of 617 days, as compared with 178 days for those aged over 85. It was suggested that the short-term morbidity of screening might outweigh the benefits in women aged over 85, but that in all other age groups the benefits outweighed the costs.

Recently King *et al.*[37] examined screening practice among women in retirement communities in Pennsylvania. A random sample was taken of 752 women aged 65 years and older, and they were asked about breast self-examination, clinical examination, and mammography. A clinical examination of the breast had been performed within the past year in 56 per cent of those aged 65–74, 54 per cent of those aged 75–79, and 48 per cent of those aged 80–84. Regular breast self-examination was performed in 44 per cent of those aged 65–74 and 37 per cent of those aged over 85. Mammograms had been performed within the previous 2 years in 51 per cent of those aged 65–74, 60 per cent of those aged 75–79, 44 per cent of those 80–84, and 24 per cent of women aged 85 years or older.

All these data are indicative of a potential role for mammographic screening in women aged over 65. Population studies on compliance and its relation to cancer education need to be carried out. The situation is in flux. The majority of developed countries are bringing in population screening, and as this becomes more widely accepted so there will be a demand from screened women to continue with regular screening mammography. Although women with severe comorbidity are unlikely to benefit from breast cancer screening, there are no grounds for exclusion of postmenopausal women from screening programs based on age alone.

Controlled randomized trials

The questions which have been raised in this chapter can be answered finally only when information becomes available from well designed controlled randomized trials. Among women of 70 years of age or more with operable breast cancer, very few prospective randomized trials have been conducted. Those which are available have examined two aspects of treatment: firstly, the role of systemic treatment as an adjuvant to standard local therapy by total mastectomy and axillary node clearance and secondly, the feasibility of replacing surgery by tamoxifen therapy.

Adjuvant therapy

Three trials of adjuvant tamoxifen therapy have been reported.[38-40] These are summarized in Table 6.4. The first, reported by the Danish Breast Cancer Cooperative Group (DBGG) included 509 women aged 70–79 who, following total mastectomy and axillary node sampling, were deemed to have 'high risk' disease, on account of involved nodes or a tumour greater than 5 cm in diameter.[40] All patients received post-operative radiation therapy to skin flaps and regional nodes and were randomized to be given tamoxifen 20 mg daily for 12 months (251 patients) or no further treatment (258 patients). Seventy-eight per cent had oestrogen receptor rich tumours.

Table 6.4 Trials of adjuvant therapy in the elderly (C control group, T treated group)

Group	Control	Treated	Relapse-free survival (%)		Overall survival (%)	
DBCG	RT	RT + Tam	C	T	C	T
	n = 258	*n* = 251	41	39	–	–
ECOG	Placebo	Tam	C	T	C	T
	n = 84	*n* = 86	52	73	77	82
BCSG	Obs	Tam + Pred	C	T	C	T
	n = 153	*n* = 167	22	36	42	49

RT, radiation therapy; Tam, tamoxifen; Pred, prednisolone.

After 7 years of follow-up no difference in relapse-free or overall survival has been observed.

This result differs from that reported by the Eastern Cooperative Oncology Group (ECOG) in which 181 women aged over 65 years were treated for node positive, oestrogen receptor positive, breast cancer by mastectomy and axillary clearance.[38] They were then randomized to receive tamoxifen 10 mg twice daily or placebo tablets.

After a median follow-up of 55 months, 73 per cent of the tamoxifen group were relapse free compared with 52 per cent of controls ($p = 0.003$). There was no significant difference in overall survival. However, the hazard rates for cancer death were 0.074 for the tamoxifen group and 0.212 for those given placebo, a ratio of 2.8 in favour of the tamoxifen treated group.

The third study, conducted by the International Breast Cancer Study Group, included 320 evaluable patients aged 65–80 years with node positive breast cancer, who following total mastectomy and axillary clearance were randomized to receive tamoxifen 20 mg and prednisolone 7.5 mg daily for 12 months (167 patients) or no systemic therapy (153 patients).[39] Relapse-free survival at 8 years was significantly less in the tamoxifen–prednisolone treated group (36 per cent versus 22 per cent $p = 0.004$) but there was no significant difference in overall survival (49 per cent versus 42 per cent).

Thus two out of these three studies have shown a significant advantage in terms of disease-free survival in elderly women given 12 months of adjuvant systemic therapy with tamoxifen; the study from Denmark has failed to do so. However, it should be noted that tamoxifen was given for only 12 months in the Danish trial. A previous study had shown that the major effect of 1 year of adjuvant tamoxifen was to reduce the number of local and regional recurrences,[41] and this would be achieved by the post-operative radiotherapy given to both arms of the trial. None of the trials has yet shown any effect on overall mortality, possibly because this may not become apparent for several years, during which time there will have been many deaths from non-cancer related causes.[42]

Tamoxifen as primary systemic therapy

The prospective randomized trials comparing primary tamoxifen therapy with the surgical treatment of operable breast cancer are summarized in Table 6.5. The results of three of these trials have recently been published.[43,44]

The first trial was conducted at St George's Hospital, London, between 1982 and 1987, the aim being to compare surgical treatment with the use of tamoxifen alone in as broad a group of elderly patients with operable breast cancer as possible.[43] Eligible cases were aged 70 years or more, all considered to be fit for surgery. Of the total number of 222 cases presenting to the unit during the accrual period, 116 were considered eligible. The main reasons for non-entry were patient refusal (22), locally inoperable disease (15), serious intercurrent illness (11), psychological unsuitability (11), or a second malignancy (10). Following cytological confirmation of the diagnosis by fine needle aspiration, eligible patients were randomized either to receive tamoxifen 20 mg daily (60 patients) or to undergo surgery (56 patients). Surgery was not standardized, but in general consisted of wide local excision for small tumours and mastectomy for large tumours. In the event 18 per cent of those with small (T1,2) and 67 per cent of those with large (T3,4) tumours were treated by mastectomy.

Results were reported after a median follow-up period of 3 years. Local progression or relapse of disease occurred in 25 per cent of the tamoxifen group compared with 38 per cent in those treated by surgery. In those treated by wide local excision alone there was a 35 per cent local relapse rate. Relapse-free and overall survival rates were identical in the two groups. The authors concluded that tamoxifen gave as effective control as surgical treatment, but this conclusion has been challenged.

A second controlled randomized trial comparing primary tamoxifen therapy with surgery has been reported from Nottingham.[44] Patients were entered between 1982 and 1987 and were all aged 70 years or more; all had presented with operable breast cancer and were considered fit for general anaesthesia. They were randomized to receive tamoxifen 20 mg twice daily (68) or to be treated by a 'wedge mastectomy' (67) by which the breast was drawn away from the chest wall and excised around its base. This operation was regarded as a more rapid surgical procedure and therefore as carrying less risk than mastectomy; there was no operative mortality.

Table 6.5 Trials of tamoxifen versus surgery in elderly women with operable breast cancer

Trial	T	S	RFS	OS
St Georges	T 20 mg	Wide excision or total mastectomy	T= S	T = S
Nottingham	T 40 mg	Wedge mastectomy	S > T	T = S
CRC	T 20 mg	Wide excision or mastectomy + T 20 mg	S > T	S > T
EORTC				
10850	T 20 mg + tumourectomy	Mod. rad. mastectomy	S > T	?
10851	T 20 mg	Mod. rad. mastectomy	S > T	?

RFS, relapse-free survival; OS, overall survival; S, surgery; T, tamoxifen; Mod rad, modified radical.

The results, after a median follow-up of 24 months, indicated a high incidence of relapse in both groups. Progressive local disease occurred in 30 per cent of those given tamoxifen and 25 per cent of those treated by wedge mastectomy alone. There were 10 deaths in the tamoxifen and 17 deaths in the mastectomy groups; a difference which did not reach statistical significance. It was concluded by the authors that because of poor local control in the patients given tamoxifen, the primary treatment of operable breast cancer in the elderly should include both 'wedge' mastectomy and tamoxifen.

In the Cancer Research Campaign (CRC) trial 381 patients aged 70 years or more with operable disease were randomized to receive tamoxifen alone (20 mg daily) or surgery, as deemed to be optimal by the participating surgeon, followed by the same dose of tamoxifen.[45]

Tamoxifen alone was given to 183 patients and 171 were treated with a combination of surgery and tamoxifen. After a median follow-up of 34 months, local recurrence or progression occurred in 22 (13 per cent) of those treated surgically compared with 51 (28 per cent) of the tamoxifen group. Distant recurrence rates and mortality were similar in both groups. A recent update of the results, presented in abstract, confirms that patients treated with tamoxifen alone are more likely to require a change in their management during a median follow-up time of 3–5 years and may have reduced survival compared with those treated by surgery and tamoxifen.[46]

Two prospective randomized trials are currently being conducted by the EORTC Breast Cancer Cooperative Group. In one trial, patients aged 70 years or more with operable breast cancer are being allocated randomly for treatment by tamoxifen 20 mg daily or by total mastectomy and axillary node clearance. In the other trial, which is being carried out in different institutions, patients are being randomized for treatment by local excision of the tumour plus tamoxifen 20 mg daily or by total mastectomy and axillary clearance. Guy's Hospital in London has been a major contributor to the second trial and a preliminary analysis has been conducted from a median follow-up of 5 years in 123 patients, 63 treated by local excision of the tumour and tamoxifen, 60 by mastectomy and axillary clearance. Local relapse rates have proven to be higher in those treated by limited surgery and tamoxifen (25 per cent) than in those treated by mastectomy (5 per cent). There has been a greater number of deaths in the former group, but this difference has not reached statistical significance.

The results of these trials point to the obvious conclusion that if rapid local control of the disease is the prime aim, this is best achieved by adequate local surgery. All three reported trials of primary tamoxifen therapy have compared this with limited surgery, such as local excision of the tumour alone, which does not help define the role of what many now be regarded as adequate local treatment (total mastectomy and node clearance, or local excision followed by radiotherapy) for primary disease.

It has therefore been assumed that less radical local treatment is justifiable in the elderly. As indicated above, this is unproven.

Conclusions

It is most important that the elderly are allowed choice and participation in decisions concerning their treatment. If the patient decides that she would like to have tamoxifen alone, the diagnosis should be confirmed cytologically, or by needle biopsy, and ERICA estimation of steroid receptor status should be performed. Provided that more than 20 per cent of the cells stain, tamoxifen is a reasonable therapy. If less than 20 per cent the patient should be offered surgical treatment.

If surgery is being carried out it should be optimal. Mastectomy should include an axillary clearance to maximize local control, and irrespective of nodal status patients should be given adjuvant tamoxifen at a dosage of 20 mg daily. It is unnecessary to give higher dosages which may be associated with increased side effects such as hot flushes and vaginal discharge.

Present techniques of breast conservation using standard fractions radiotherapy may be inappropriate for many elderly patients. This may be circumvented with the use of high dose implants,[47] or the administration of low dose radiation by high doses in a small number of well spaced fractions.[48]

It is important that clinical trials of treatment do not exclude patients on grounds of age alone, and that these trials are well designed and conducted. Only in this way will there be improvements in treatment for breast cancer in the elderly.

Key points

- Approximately 40 per cent of breast cancers occur in women aged 70 years or older.
- Breast cancer is not a less aggressive disease in older women.
- Mammographic screening of women aged over 65 years needs evaluation.
- Adjuvant tamoxifen should be given to the majority of elderly patients after adequate primary treatment.
- Controlled trials have shown that better local control is achieved after mastectomy compared with tamoxifen alone. This may also have an impact on mortality.
- Primary treatment with tamoxifen should be used only in patients with primary tumours containing more than 20 per cent of cells with oestrogen receptor protein.

References

1. Office of Population Census and Surveys. (1990). Cancer statistics registration: cases of diagnosed cancer registered in England and Wales 1985. *London Cancer Fact Sheet*, series MBI No. 18. HMSO, London.
2. White, E., Lee, C.Y., and Kristal, A-R. (1990). Evaluation of the increase in breast cancer incidence in relation to mammography use. *J. Natl. Cancer Inst.*, **82**, 1546–52.
3. Adami, H-O., Walker, B., Rutqvist, L.E., Persson, T., and Ries, L. (1986). Temporal trends in breast cancer survival in Sweden: significant improvement in 20 years. *J. Natl. Cancer Inst.*, **76**, 653–9.
4. Day, N. and Chamberlain, J. (1988). Screening for breast cancer: Workshop report. *Eur. J. Cancer Clin. Oncol.*, **24**, 55–9.
5. Brinkley, D. and Haybittle, J.L. The curability of breast cancer. *Lancet*, **ii**, 95–6.
6. Early Breast Cancer Trialists Collaborative Group (1992). Systemic treatment of early breast cancer by hormonal, cytotoxic or immune therapy – 133 randomised trials involving 31 000 recurrences and 24 000 deaths among 75 000 women. *Lancet*, **339**, 1–14.
7. Forrest, A.P.M. (1990). Breast cancer: the decision to screen. Queen Elizabeth The Queen Mother Fellowship. Nuffield Provincial Hospitals Trust.
8. Mueller, C.B., Ames, F., and Anderson, G.P. (1978). Breast cancer in 3558 women, age as a significant determinant in the rate of dying and causes of death. *Surgery*, **83**, 123–32.
9. Adami, H-O., Walker, B., Holmberg, L., Persson, I., and Stone, B. (1986). The relation between survival and age at diagnosis of breast cancer. *N. Engl. J. Med.*, **315**, 559–63.
10. Fentiman, I.S. (1990). *Detection and treatment of early breast cancer*, pp. 193–206. Martin Dunitz, London.
11. Chryssos, A.E. and Bondi, R.P. (1984). Breast cancer in women over the age of 80. *Breast*, **10**, 13–15.
12. Robins, R.E. and Lee, D. (1985). Carcinoma of the breast in women 80 years of age and older: still a lethal disease. *Am. J. Surg.*, **150**, 655–8.
13. Davis, S.J., Karrer, F.W., and Moor, B.J. (1985). Characteristics of breast cancer in women over 80 years of age. *Am. J. Surg.*, **150**, 655–8.
14. Host, H. (1986). Age as a prognostic factor in breast cancer. *Cancer*, **57**, 2217–21.
15. Berg, J.W. and Robbins, G.F. (1961). Modified mastectomy for older, poor risk patients. *Surg. Gynecol. Obstet.*, **113**, 631–4.
16. Kraft, R.O. and Block, G.E. (1962). Mammary carcinoma in the aged patient. *Ann. Surg.*, **156**, 981–5.
17. Cortese, A.F. and Cornell, G.N. (1975). Radical mastectomy in the aged female. *J.Am. Geriat. Soc.*, **23**, 337–42.
18. Kesseler, H.J. and Seton, J.Z. (1978). The treatment of operable breast cancer in the elderly female. *Am. J. Surg.*, **135**, 664–6.
19. Nemoto, T., Vang, J., and Bedwani, R.N. (1980). Management and survival of female breast cancer: results of a national survey by the American College of Surgeons. *Cancer*, **45**, 2917–24.
20. Schaefer, G., Rosen, P.P., Lesser, M.L., Kinne, D.W., and Beattie, E.J. (1984). Breast carcinoma in elderly women: pathology, prognosis and survival. *Pathol. Annu.*, **19**, 195–219.

21. Kantorowitz, D.A., Poulter, C.A., Sischy, B., Paterson, E., Sobel, S.H., Rubin, P., Dvoreisky, P.A., Mishalak, W., and Doane, K.L. (1988) Treatment of breast cancer among elderly women with segmental mastectomy or segmental mastectomy plus postoperative radiotherapy. *Int. J. Radiat. Oncol. Biol. Phys.*, **15**, 263–70.

22. Reed, M.W.R. and Morrison, J.M. (1989). Wide local excision as the sole primary treatment in elderly patients with carcinoma of the breast. *Br. J. Surg.*, **76**, 898–900.

23. Helleberg, A., Lundgren, B., Norin, T., and Sander, S. (1982). Treatment of early localised breast cancer in elderly patients by tamoxifen. *Br. J. Radiol.*, **55**, 5111–13.

24. Bradbeer, J.W. and Kingdon, J. (1983). Primary treatment of breast cancer in elderly patients by tamoxifen. *Clin. Oncol.*, **9**, 31–4.

25. Akhtar, S.S., Allan, S.G., Rodger, A., Chetty, U., Smyth, J.S., and Leonard, R.C.F. (1991). A ten year experience of tamoxifen as primary treatment of breast cancer in 100 elderly and frail patients. *Eur. J. Surg. Oncol.*, **17**, 30–5.

26. Horobin, J.M., Preece, P.E., Dewar, J.A., Wood, R.A.B, and Cuschieri, A. (1991). Long-term follow-up of elderly patients with loco-regional breast cancer treated with tamoxifen alone. *Br. J. Surg.*, **78**, 213–17.

27. Anderson, E.D.C., Forrest, A.P.M., Hawkins, R.A., Anderson, T.J., Leonard, R.C.F, and Chetty, U. (1991). Primary systemic therapy for operable breast cancer. *Br. J. Cancer*, **63**, 561–6.

28. Gaskell, D.J., Forrest, A.P.M., de Carteret, S., Chetty, U., Sangster, K., and Hawkins, R.A. (1992). Use of primary tamoxifen therapy for elderly women with breast cancer. *Br. J. Surg.*, **79**, 1317–20.

29. McCarty, K.S., Silva, J.S., Cox, E.B., Leight, G.S., Wells, S.A., and McCarty, K.S. (1983). Relationship of age and menopausal status to estrogen receptor content in primary carcinoma of the breast. *Ann. Surg.*, **197**, 123–7.

30. Gaskell, D.J., Hawkins, R.A., Sangster, K., Chetty, U., Forrest, A.P.M. (1989). Relation between immunocytochemical estimation of oestrogen receptors in elderly patients with primary operable breast cancer in response to tamoxifen. *Lancet*, **i**, 1044–6.

31. Coombes, R.C., Powles, T.J., Berger, U., Wilson, P., McClelland, R.A., Gazet, J.-C., Trott, P.A., and Ford, H.T. (1987). Prediction of endocrine response in breast cancer by immunocytochemical detection of oestrogen receptor in fine needle aspirates. *Lancet*, **ii**, 701–3.

32. Clinical Practice Committee American Geriatrics Society (1989). Screening for breast cancer in elderly women. *J.Am. Geriat. Soc.*, **37**, 883–4.

33. Hobbs, P., Kay, C., Friedman, E.H.I., St Leger, A.S., Lambert, C., Boggis, C.R.M., Howard, T.M., and Asbury, D.L. Response by women aged 65–79 to invitation for screening for breast cancer by mammography: a pilot study. *Br. Med. J.*, **301**, 1314–16.

34. Harris, R.P., Fletcher, S.W., Gonzalez, J.J., Lannin, D.R., Degnan, D., Earp, J.A., and Clark, R. (1991). Mammography and age: are we targeting the wrong women? *Cancer*, **67**, 2010–14.

35. Weinberger, M., Saunders, A.F., Samsa, G.P., Bearon, L.B., Gold, D.T., Brown, J.T., Booher, P., and Loehrer, P.J. (1991). Breast cancer screening in older women practices and barriers reported by primary care physicians. *J. Am. Geriat. Soc.*, **39**, 22–9.

36. Mandelblatt, J.S., Wheat, M.E., Monane, M., Moshief, R.D., Hollenberg,

J.P., and Tang, J. (1992). Breast cancer screening for elderly women with and without comorbid conditions. *Ann. Intern. Med.*, **116**, 722–30.

37. King, E.S., Resch, N., Rimer, B., Lerman, C., Boyce, A., and McGovern-Gorchov, P. (1993). Breast cancer screening practices among retirement community women. *Prev. Med.* **22**, 1–19.

38. Cummings, F.J., Gray, R., Davis, T.E., Tormey, D.C., Harris, J.E., Falkson, G., and Arseneau, J. (1986). Tamoxifen versus placebo: double blind adjuvant trial in elderly women with stage II breast cancer. *Natl. Cancer Inst. Monogr.*, **1**, 119–23.

39. Goldhirsch, A., Gelman, R.S., Gelber, R.D., and Castiglione, M. (1990). Treatment for breast cancer in elderly patients. *Lancet*, **336**, 564.

40. Mouridsen, H.T., Andersen, A.P., Brinker, H., Dombernowsky, P., Rose, C., and Andersen, K.W. (1986). Adjuvant tamoxifen in postmenopausal high-risk patients: present status of Danish Breast Cancer Cooperative Group Trials. *Natl. Cancer Inst. Monogr.*, **1**, 115–18.

41. Goldhirsch, A., Gelber, R.D., Tattersall, M.N.V., Rudenstam, C.-M., and Cavalli, F. for the Ludwig Breast Cancer Study Group (1985). Endocrine adjuvant therapy for breast cancer. *Lancet*, **i**, 1274.

42. Castiglione, M., Gelber, R.D., and Goldhirsch, A. (1990). Adjuvant systemic therapy for breast cancer in the elderly: competing causes of mortality. *J. Clin. Oncol.*, **8**, 519–26.

43. Gazet, J.-C., Markopoulos, C., Ford, H.T., Coombes, R.C., Bland, J.M., and Dixon, R.C. (1988). Prospective randomised trial of tamoxifen versus surgery in elderly patients with breast cancer. *Lancet*, **i**, 679–81.

44. Robertson, J.F.R., Todd, J.H., Ellis, I.O., Elston C.W., and Blamey, R.W. (1988). Comparison of mastectomy with tamoxifen for treating elderly patients with operable breast cancer. *Br. Med. J.*, **297**, 511–14.

45. Bates, T., Riley, D.L., Houghton, J., Fallowfield, L., and Baum, M. (1991). Breast cancer in elderly women: a Cancer Research Campaign trial comparing treatment with tamoxifen and optimal surgery with tamoxifen alone. *Br. J. Surg.*, **78**, 591–4.

46. Bates, T., Riley, D., Houghton, J., and Baum, M. (1992). *Is tamoxifen adequate treatment for breast cancer in elderly patients; is there still an issue.* Surgical Research Society, London.

47. Fentiman, I.S., Poole, C., Tong, D., Winter, P.J., Mayles, H.M.O., Turner, P., Chaudary, M.A., and Rubens, R.D. (1991). Iridium implant treatment without external radiotherapy for operable breast cancer: a pilot study. *Eur. J. Cancer*, **27**, 447–50.

48. Ashby, M.A., Campana, F., Fourquet, A., Julien, D., and Vilcoq, J.R. (1988). Management of breast cancer in the elderly. *Lancet*, **2**, 461–2.

Acknowledgements

We are grateful to Mrs Dorothy May and Mrs Kim Bullock for secretarial assistance. A.P.M.F. was in receipt of a Leverhulme Emeritus Fellowship.

7

Prostate cancer

N.J.R. George

Introduction

Prostate cancer is escalating rapidly into a major cause of medical, pathological, and economic concern. Estimated 15 years ago to be the fifth most
frequent cancer in men, with nearly 200 000 cases diagnosed worldwide,[1] it
has recently been suggested that it may be the most common malignancy
in humans.[2] Whilst in England and Wales the disease was diagnosied in
5700 patients and was the fourth most common male cancer in 1970, more
recent figures indicate it has become the second most common cause of
male cancer deaths (10 per cent of all male neoplasms) accounting for
approximately 7500 deaths in 1988. Within the UK as a whole new case
registration has increased rapidly in recent years, 10 820 patients being
initially diagnosed in 1985.[3,4] In the USA newly diagnosed cases of clinical
prostate cancer 20 years ago were 42 000[5] and have increased to over
100 000 in 1990;[6-8] associated annual deaths from the disease escalated
from 14 000 to 28 500 within the same period.[5,9] The disease is now the
most commonly diagnosed cancer in elderly US males.[10]

Epidemiological and pathological considerations

Demographic trends

Worldwide, it is now well established that there are marked variations
in the incidence (i.e. histologically proven disease in patients who are
alive) of prostate cancer with the highest rates apparent in non-whites
in the USA and the lowest in oriental races.[11,12] However, these overall
statistics conceal figures of considerable epidemiological interest. Rates for
various ethnic groups located within similar geographical zones may differ
considerably;[13] conversely, amongst Scandinavians histological proof of
disease is obtained twice as often in Swedish as opposed to Danish
citizens,[12] as shown in Table 7.1.

Table 7.1 Incidence of prostate cancer: age standardized rates per 100 000 population (world standard population) based on data from ref. 14

Population	Incidence
San Francisco Bay area	
Black	92.2
White	47.4
Chinese	18.6
Japanese	12.7
Hawaii	
White	59.7
Hawaiian	42.5
Japanese	35.9
Chinese	25.8
Scandinavia	
Sweden	44.4
Norway	38.9
Finland	27.2
Denmark	23.6
New Zealand	
Maori	39.8
Non-Maori	30.7
Singapore	
Indian	6.7
Chinese	4.8
China	
Shanghai	0.8

Mortality from prostate cancer also demonstrates wide international variation.[12,15] Rates in the West Indies are very high as are those in Sweden and Norway; the lowest levels are found in the Far East, Thailand, Hong Kong, and Japan. Racial differences may be pronounced; within the continental USA mortality for blacks is observed to be approximately twice that for whites.[16,17] Whether these differences relate to genetic variation in the disease process or to differing patterns of health care delivery remains unclear. The differing life expectations of US and West African blacks means that comparative studies have to be interpreted with care.[18]

Prostate cancer is rare under 50 years[19] but subsequently the clinically diagnosed disease incidence increases more rapidly than that for other tumours as age progresses,[14,20] a forty-fold increase in incidence being observed by age 85.[14,21] Clearly, the changing demographic profile of many western societies may have a major potential impact on prostate cancer health care demands. In the USA it is estimated that the male population in excess of 65 years will increase by 64 per cent from 8.4 million in 1970 to 13.8 million by the turn of the century.[22] The future healthcare burden in such an elderly group will vary according to the thoroughness of clinical examination, the intensity of preoperative investigation, the prevalence of surgery for prostatic disorders, and the sophistication of pathological processing of material obtained at operation.

Superimposed upon, and dwarfing, the problem of clinical prostate cancer (both medically and economically) is the issue of disease diagnosed after death – latent prostate cancer. Despite doubts concerning definition there is no doubt that this form of disease is very common.[23–25] Variations observed in the natural history of prostate cancer within differing populations subject to separate environmental influences allow a unique opportunity to study patterns of disease development. The biological and clinicopathological relationship between 'latent' and clinically manifest disease has presently become the focus of intense scientific and research endeavour.

Latent prostate cancer

Latent cancer[23,25] is found by chance at autopsy, the patient having another certified cause of death. A critical examination of the meaning of 'latent' disease may be instructive, as in such cases ante-mortem evidence regarding the presence of clinical disease is often circumstantial. Phosphatase estimations, bone scans, or skeletal surveys are rarely requested in the elderly if the diagnosis is not suspected as skeletal discomfort is far from rare in old age. Observations on iliac and para-aortic nodes and marrow from lumbar vertebrae (if sampled) are often unreported in most retrospective autopsy studies which restrict themselves to the gland itself with attached seminal vesicles.[24,26] Clifton and associates reported on 49 patients with histological proof of disease before bone scans were available. Although skeletal surveys were normal, 9 per cent had tumour cells in bone marrow aspirates whilst 66 per cent were pain free.[27] It may be concluded that in patients without histological proof of prostate cancer, lack of pain and normal radiographs do not ensure the absence of metastatic tumour. Hence the widespread concept of latent cancer as a 'benign' disorder[23] rests on the premise that the physician will not seek the diagnosis in asymptomatic patients. No doubt in the majority of cases—bearing in mind the small volume and low grade of most latent tumours—such assumptions may be correct (see below) but evidence from retrospective autopsy studies remains circumstantial.

Despite these reservations, autopsy studies agree that histological evidence of prostate cancer in patients who are certified as dying from other causes is increasingly common as age progresses.[24,26,28] More intensive pathological specimen examination yields more positive histological findings and whilst early studies reported on one or occasionally two sections of the gland,[29] later work examining 4 mm step sections identified latent changes in 41 and 57 per cent of patients over 70 and 80 years respectively,[25] figures confirmed by other authors.[24,28]

Although the prevalence of latent cancer in the population is often reported as being similar worldwide, geographic variation in the frequency of the observation depends to a degree on histological interpretation and age profile of censored autopsies.[30-32] 'Small' carcinomas occurred in approximately 12 per cent of autopsies from each of seven geographical locations but the assessment by octant count led to some pathological disagreement. 'Medium' and 'large' carcinomas (occupying more octants) appeared to be related to countries with a high clinical incidence although it was notable that in Sweden more than twice the number of autopsies reported (high frequency of larger carcinomas) were performed in patients over 75 years old—44 per cent as compared with 19 per cent in Hong Kong and 8 per cent in Uganda.[30] Hence 'large latent' cancers might be evolving clinical cancer or might merely be a function of patient longevity. Additionally, conscientious public health policies might be responsible for the high rates noted in Sweden. In Malmö 99 per cent of patients dying in hospital (65 per cent of all urban deaths) are autopsied and the prostate examined by 5 mm step section.[28] Such a detailed examination policy is likely to give a very exact profile of any particular disease pattern.

Clinical disease

Unlike the frequency of latent cancer, the incidence of clinically diagnosed tumour varies widely amongst different international groups (Table 7.1). Age standardized rates for Alameda black patients are almost 100 times greater than those for Chinese patients in Shanghai.[12] However, whilst it is now estimated that there may be 10 million US males with latent disease,[14] a small, but rapidly increasing, percentage is identified annually. The increasing conversion of this 'pool' of latent disease to clinically detected cancer might be explained in part by increasing use of transurethral resection in an ageing population.[33] Furthermore, thorough pathological search of resected material in centres with an active interest in prostate diseases might, as with autopsy studies, be expected to reveal increasing amounts of hitherto unsuspected (T0, stage A) cancer. Chisholm and coworkers discovered approximately 25 per cent of new patients undergoing operation for supposed benign disease exhibited neoplastic change.[34]

Nevertheless, the differing patterns of disease in eastern and western older males cannot be explained merely by variations in clinical or pathological practice. Life expectancy for Japanese men is approximately the same as that for US men yet the age adjusted incidence of cancer is over 15 times greater in the latter population.[35] Migratory studies of Japanese men settling in Hawaii or California [15,36] reveal a marked increase in clinical cancer, emphasizing that environmental or other epidemiological aspects of the disease are of greater importance than genetic factors.

Conversely, reports of cancer in male relatives of patients,[37] the worldwide predisposition of black races to the disease, and the observation that ethnic groups in the same areas do not enjoy the same incidence,[12] imply that genetic influences on disease expression cannot be totally discounted. Diet,[38] body weight,[39] and physical and sexual activity[40] are other factors which have been intensively studied in differing population groups in an attempt to identify factors responsible for conversion of histologically silent to biologically active disease. Examination of the foregoing data concerning the 'pool' of latent histologically identifiable lesions and the variable worldwide clinical expression of the disease serves to illustrate that there can be no unified concept of the 'natural history of prostate cancer'. Thus Scandinavians, British, affluent WASPs, poor North American blacks, new wealthy Hong Kong Vancouver immigrants, and third generation Japanese in California will all be subject to a large number of variable influences. Evidence obtained from one cohort regarding natural history may not be of assistance in assessing the prognosis of others.

Determinants of disease progression

Localized disease

In any single population the natural history of disease confirmed histologically during life is best observed when no treatment is offered which could alter the course of the disorder. Older men in the UK of average age 72 years who presented with symptoms of outflow tract obstruction due to cancer localized to the gland showed 84 per cent T stage progression over a 6 year period but few died of the cancer itself.[41] Similar results were reported from Scandinavia,[42,43] and there is general agreement that in such older men the risk of dying from prostate cancer depends essentially on whether or not they will succumb to other forms of fatal disease before the tumour metastasizes. Generally, males in North America present much earlier or are initially seen when asymptomatic at screening or 'well man' clinics rendering comparison with such observations from the UK and Scandinavia difficult.

Most workers agree that tumour grade[44,45] and volume[46-48] are important prognostic indicators for prostate cancer and this view naturally leads to the assumption that most autopsy-identified 'latent' tumours (being usually small volume low grade lesions) will not have spread beyond the confines of the gland. Nevertheless, such concepts of tumour spread— local enlargement followed by metastases to adjacent sites[49]—are almost certainly an oversimplification of the metastatic process (see below). Evidence from surveillance studies shows that, whilst the local tumour may be palpably enlarging at each review, metastases in bone take some years to develop.[41] The significance of high volume, low grade tumour has been questioned.[50] Occasional cases of widespread metastases in which the primary was undetectable by any technique prior to autopsy (T0 Nx M1) can be recalled by most surgeons. Survival in a subset of patients presenting with massive lymphadenopathy yet negative bone scans was recently described by Sandhu and associates. Significantly enhanced survival (70 per cent at 5 years) was noted when compared with those with positive bone scans at presentation (20 per cent); histological examination of primary or nodal tissue failed to identify any pattern linked to the unusual metastatic preference of the tumour.[51] Thus, whilst it is undoubtedly true that histological grade and tumour volume predict prognosis in general, there are enough exceptions to cause doubt in any one individual patient seeking advice from his physician.

Microscopic cancer will be detected in approximately 10–15 per cent of patients undergoing resection for presumed benign prostatic hyperplasia, the incidence rising with the age of the patient and pathological sectioning technique. Debate continues as to the prognostic significance of stage A tumours. Epstein *et al.*[52] quoting follow-up of an earlier series,[53] originally detected progression in only 1 out of 49 patients with focal change as compared with 11 out of 33 patients with diffuse disease. Longer term, however, although a significant number died from other causes which thus masked the true progression rate, 16 per cent (eight patients) had advancing disease leading to death in six. In these cases neither original volume or grade of tumour predicted progression.[52] Such outcomes have encouraged radical prostatectomy to be considered even in cases of focal low grade tumour.[54]

Whilst the reasons for such advice are quite understandable, the medical and politico-economic impact of the decision has to be faced. Apart from the major costs involved, an over-optimistic view of treatment efficacy in cases with lesions of low biopotential may be encouraged. Just as the introduction of the Franzen aspiration cytology technique in Sweden resulted in the incidence of prostate cancer doubling between 1958 and 1971[55] with resulting apparent treatment benefits[56,57] (probably owing to inclusion of patients with lesions of questionable biopotential), so radical treatment for prostatic lesions of low biopotential may distort clinical judgement. Recent advice suggests that it is probably best at the present time to suppress emotional arguments in favour of informed and objective endeavour to understand better the natural history of early prostate cancer.[58]

Although a uniform pathological definition of A2 diffuse disease remains elusive, there is no doubt concerning the threat posed by these lesions discovered at endoscopy.[53,59] If the entire gland is not available for examination 'small', 'focal' or 'percentage involvement' are indefinite quantities, and a number of authors have debated the role of repeat resection.[60-62] The recent demonstration of heterogenic multifocal and multigrade tumours—well recognized in patients with more advanced local lesions[63]—in radical prostatectomy specimens removed for single focus, low grade, low volume disease,[54] further hinders the attempt to distinguish between these two types of disease.

Metastatic disease

In contrast to localized disease there is general agreement that the outlook for patients who present with or develop bony metastases is poor. Until very recent observations on the efficacy of combined treatment became available (see below), results of hormone manipulation studies showed remarkable unanimity over the half century since Huggins' landmark observations; approximately 50 per cent of all patients will not survive 2 years and 5 year survival is between 15 and 20 per cent.[64] Following hormonal manipulation, approximately 75–80 per cent of patients respond to androgen deprivation but relapse occurs in the majority after a median time of 12–15 months. Clinical and experimental observations strongly suggest that relapse is related to tumour heterogeneity with respect to androgen sensitivity,[65] though whether such variability arises as a function of postmanipulative adaptive transformation or fundamental clonal hetrogeneity remains open to question.[66]

Patients with relapsed disease may be temporarily relieved by a number of short-term measures (see below) but in general the subsequent outlook remains extremely poor. Overall, it has been long observed that patients presenting with metastatic disease but good general performance scores at diagnosis fare better than their more debilitated counterparts despite apparently similar tumour burden; the scientific basis for this credible observation has not been precisely defined.

Investigation of prostate cancer in the elderly

The physician investigating the elderly man must bear in mind the often taxing nature of the many scans and radiographs which are presently available for accurately staging prostate cancer. Insistence in every case on the full range of investigations including ultrasound guided biopsy, computerized tomography, and bone scanning, which will be performed prior to the possible need for endoscopy and transurethral resection, may well place a burden on the patient and his relatives which will result in reluctant cooperation during future attempts to monitor disease activity. In the elderly especially, respect for quality of life is likely to be most beneficial to patients, relatives, and doctor alike.

Performance status

As mentioned above, performance status (together with tumour grade) is one of the most powerful prognostic indicators in any one patient with prostate cancer[67] and as such it is helpful to make a detailed note of this status at the commencement of treatment, even in the non-trial environment.

Digital rectal examination

Although the limitations of digital rectal examination (DRE) with regard to tumour staging are well described,[68] every patient demands a rectal examination by a competent physician with accurate recording of findings for comparison with future observations. The common finding (in the elderly) of a T2–T4 prostate tumour will cause little difficulty. Often, however, suspicious general symptoms and signs—back pain, sclerotic areas on pelvic radiograph – may be accompanied by an apparently normal or near normal rectal examination and it is recognized that small volume primary lesions (less than 1.5 cm diameter) may be detected digitally in only a minority of patients.[69] Under these conditions the need for further investigation of the gland by transrectal ultrasound will depend on the presence or otherwise of outflow tract symptomatology as well as biochemical or radiological evidence of metastatic disease.

Screening for prostate cancer in an elderly population

Screening for prostate cancer remains a controversial subject at the time of writing.[70] To date regular DRE programmes have not been in place long enough to show a reduction in disease-specific mortality rate and by definition will be unable to detect small tumours which are most amenable to cure. In the elderly especially, the question of biological variability may mean that pathologically detectable cancer will not necessarily equate with clinically significant disease. For these reasons it is reasonable at the present time to advise against routine rectal examination for prostatic abnormality in the absence of general clinical disease or localized urological dysfunction.

Nevertheless, this policy assumes a reasonable level of patient awareness with regard to the meaning and significance of 'prostatic' symptoms; levels of biological education vary widely between different populations.

Transrectal ultrasound

Elderly patients without symptoms of outflow tract obstruction who will not necessarily require endoscopy and transurethral resection (which will provide the histological evidence required before therapy can commence) may conveniently be submitted to transrectal ultrasound with synchronous biopsy. Recent advances in scanner technology and software have resulted in excellent real time image formation with extremely accurate biopsy capability. However, the technique demands a high level of skill and experience to interpret the images obtained and ensure representative tissue is sampled for histological analysis.[71] For these reasons the test is unlikely to be requested in the majority of patients with unequivocal digital evidence of cancer in whom treatment can be initiated on the basis of results obtained by other means.

Acid phosphatase

The reputation of acid phosphatase has suffered severely in recent years as a result of work on the epithelial marker prostate specific antigen (PSA) identified in 1979. Nevertheless, and particularly in the elderly population, the test deserves consideration as a base-line assay that may rapidly assist the clinician in the assessment of the general state and prognosis of the case in hand.

Assays for prostatic acid phosphatase are generally cheap and give reliable results in laboratories with good quality control programmes.[72] The marker is not raised in 15–20 per cent of cases, usually with high grade tumours, in which it is assumed that the enzyme cannot be expressed by the anaplastic cell lines. It is also not raised in disease localized to the gland but elevates with spread to nodes or bone, often in proportion to disease burden. As 55 per cent of patients within typical UK health districts present with metastatic disease to bone,[41] it can be seen that this cheap assay represents a very cost effective marker for out-patient surveillance of patient progress. Clearly, it cannot be used for observational studies of patients with localized disease. Although on treatment relapse patients with metastatic disease tend to demonstrate rising PSA levels before acid phosphatase levels, the benefit of this foreknowledge as regards therapeutic options remains unclear (see below). In summary, and with the provisions mentioned above, acid phosphatase may provide a cost effective answer for those physicians wishing to monitor routine patients outside of clinical trials with either treated or untreated metastatic disease.

Prostate specific antigen

Isolated in 1979,[73] prostate specific antigen has replaced acid phosphatase in many laboratories owing to its increased usefulness in the assessment and monitoring of patients with both localized and metastatic prostate cancer. PSA is tissue specific rather than cancer specific and as such considerable overlap exists between patients with benign prostatic hypertrophy and localized neoplastic disease.[72] Levels rise with T stage but considerable overlap occurs between stages. The level of PSA elaborated from benign tissue is, weight for weight, less than that elaborated from malignant tissue[74] and hence much effort has been extended in an attempt to relate PSA levels to gland volumes in order to obtain a separation between benign and malignant disease.[75]

Prostate specific antigen is particularly useful when judging the adequacy of resection following radical prostatectomy (the level should be zero) and may be an indication of good prognosis when values fall to less than 10 ng/ml within 6 months of hormone manipulation for metastatic disease.[76] In the elderly patient PSA is very helpful for monitoring progress in those patients with 'localized' disease (bone scan negative) in whom age or other considerations (grade, stage) determine that a surveillance policy is appropriate.[41] In such cases the 6 monthly levels of PSA provide a sensitive marker of disease activity and may warn the physician long before bone metastases occur that treatment should be offered.[72] In general, in such patients the rate of change of PSA (if any) is of much greater clinical use than a 'single shot' estimation which, except in relatively advanced cases, gives little information about the biological potential of the tumour.

Bone scintigraphy

Bone scanning with complementary plain films as necessary is mandatory in all cases with histological proven prostate cancer. It is now recognized that the presence or absence of bony changes at the time of diagnosis[41,43] has a major influence on patient survival, regardless of the treatment subsequently offered. For patients with 'localized' disease with an initially negative scan, estimation of PSA has largely replaced regular follow-up scintigraphy unless significant elevation of the marker is detected. This policy is less troublesome for both the patient and his family (who usually have to wait during the lengthy radioisotope process) and carries the additional benefit of a major reduction in surveillance costs.

Prediction of metastatic biopotential

Although local growth of prostate cancer is a significant cause of urological morbidity, it is the capacity of the cells to disseminate to other sites that almost invariably leads to the death of the patient. Local invasion and release from the primary focus, followed by a distant attachment, implantation, angiogenesis, and multiplication,[77] are recognized as some of the many stages in the complex metastatic process.

Whilst, as noted above, most investigators believe that tumour grade and volume are the primary determinants of the likelihood of spread, it is becoming increasingly clear that the presence of histologically verifiable prostate cancer cells does not *per se* guarantee that all criteria for malignant invasion have been met. Whether the 'malignant' cells so identified lack the ability to invade in such cases or whether such cells are able to reach the circulation and marrow spaces but unable to seed remains unknown at the present time. Biological characteristics which facilitate or retard the metastatic process are presently the subject of intense investigation.

Early studies identified malignant cells in the peripheral blood of patients undergoing prostatic massage or transurethral resection.[78] Such procedures are often undertaken in prostate cancer patients presenting with obstructive symptoms, and in those subsequently followed by an observational policy, bony metastatic disease is slow to develop and does not appear to be an immediate consequence of the operation itself. Other authors have confirmed that, whilst the possibility of malignant emboli circulating during the procedure cannot be discounted, distant metastatic disease is not a consequence of needle biopsy or transurethral resection.[79,80] Bearing in mind the vascular spaces open during prostatic resection and the large volume of malignant cells circulating in the irrigant solution, it would be surprising if tumour embolism did not occur. That distant implantation does not apparently take place might reflect either inadequate tumour biopotential or competent (possibly enhanced) host defence mechanisms. Whether cells escape from the primary growth as a result of iatrogenic manipulation, production of proteolytic enzymes such as collagenases,[81,82] decreased cellular adhesion,[83] or other release mechanisms, survival at the distant metastatic locus will depend on the ability of the cellular embolus to adapt to the host environment.

Expression of proteolytic enzymes[83] enables tumour access to host stroma by penetration of basement membrane, whilst substrate attachment may be facilitated by adhesive cell surface glycoproteins and fibronectins.[84] Study of cells with high and low metastatic ability showed low invasive potential to be associated with negative surface charge.[85] Oligosaccharide profiles have been studied and compared in both primary and associated secondary metastatic nodal material.[86,87]

Individual tumour systems have also been examined in terms of transferrin receptor content, tumour ploidy, and active cell motility. Increased transferrin receptor levels were found in DU 145 cell lines but not in stromal cell fractions derived from benign prostatic tissues.[88] Non-diploid tumours are generally more aggressive,[89–91] although debate continues as to the relative merits of flow cytometric analysis and histopathological grading.[92] Increased motility and metastatic potential has been observed following transfection of Dunning tumour cell lines of low metastatic ability with the ras oncogene.[93,94]

A number of workers have investigated the possibility that metastatic implantation might be promoted by the host skeletal microenvironment. Manishen and associates observed that tumour growth was stimulated by products of bone resorption,[95] whilst unidirectional migration of tumour cells was noted in response to exposure to collagen fragments[96] and attachment of tumour cell aggregates was enhanced by bone resorption products *in vitro*.[97] Bone marrow stromal cells have been shown to stimulate the growth of human tumour cells[98] including prostate,[99] an effect that might explain the prediliction of some prostatic cells to seed in bone.

Although most effort has been spent on examining the biocapability of the primary tumour, it is clear from these experiments that all aspects of the host–tumour relationship must be examined if the final expression of the metastatic process is to be understood. Future studies into the natural history of early prostate cancer will of necessity address tumour cell characteristics in the context of their own skeletal microenvironment.

Treatment of prostate cancer in the elderly

The diagnosis of prostate cancer is increasing rapidly in most centres in the western world,[4,10] as a result of both heightened patient awareness and more thorough pathological processing of clinical material. Expectations of old age both in terms of years and quality of life vary widely between countries and between the socio-economic groups contained within those countries. It is unfortunately often the case that those groups with the greatest prostate cancer problem come from the poorest sections of society where expectations and provision of health care are at their lowest. Nevertheless, in general terms, most men are now living considerably longer than their grandfathers and judgements about diseases with long biological timespans such as prostate cancer have to be made with these points in mind. Neoplastic change diagnosed in an otherwise healthy 65 year old man now demands that when treatment options are considered full weight be given to the possibility of his surviving 15 or more years.

Relief of outflow tract obstruction Within the UK—as opposed to other centres in Europe and especially the USA—the great majority of patients (more than 95 per cent) later found to have prostate cancer present to their primary physician with symptoms of either outflow tract obstruction or bladder irritation. Thus before any consideration is given to anticancer treatment, surgical relief of the prostatic obstruction will be required. However, resection need not always be over radical as, if hormone manipulation is required later, medical cytoreduction will treat the disease more precisely, particularly if the sphincter muscles are involved in the pathological process.

Localized disease

Traditionally patients with localized disease (bone scan negative) have been subjected to a surveillance policy or offered radiotherapy to the gland and/or pelvic side walls. In recent years however radical retropubic prostatectomy has come to be seen as the 'gold standard' of ablative treatment which alone can offer total eradication of tumour cells.

Surveillance studies[41–43]   It is now well established from observational studies in both Scandinavia and the UK that certain tumours with low biopotential may grow extremely slowly and may not have a significant impact on patient survival. Nevertheless, the concept of 'latent' tumour as described by Franks[23] has not been seen in these studies which, when critically analysed, show a high proportion with locally advancing (albeit slowly) disease. Death due to prostate cancer is reported rarely in such series as mortality from other causes is frequent; hence it cannot be absolutely proven or assumed that the prostate cancer would not have led to the demise of the patient had it not been for the intercurrent (cardiovascular, neurological) cause of death.

Clearly therefore, the optimum treatment for any one patient has to be judged not only on the local tumour factors but on the patient's general health, well-being and life expectancy at the time of diagnosis. Overall, patients with low grade tumours of small volume (T0–T2) would seem to be a very low risk for progression to bony metastatic disease within their natural life-span, even if this is calculated at 15–20 years, and under these conditions careful surveillance would seem to be a reasonable course of action. Moderate grade tumours and those of larger volume will demand a careful balance of probabilities from the physician; under these circumstances PSA (see above) may be of great help in the less fit patient by giving an indication of epithelial activity and thus warning of the need for more immediate therapy.

Radiotherapy   Radiation therapy for localized disease has been described for many years[100] and within the UK in particular has been until recently the standard treatment for localized disease in the patient thought to be at risk because of either stage, grade, or probable life-span. A number of reports have clouded the reputation of the therapy[101,102] and local control of disease is clearly called into question in these studies. It has to be admitted however that many cases of prostate cancer sent for X-ray treatment are by definition poorly staged and certainly within the UK it remains very difficult to interpret post-irradiation survival data. Despite anxieties regarding the efficacy of this form of treatment it seems likely that X-ray treatment will remain an option for those elderly patients with unfavourable prognostic indicators who are considered to be too old or unfit for more radical therapy.

Radical prostatectomy   Whereas it is difficult to fault the logic of the proponents of radical surgery[103] in the younger patient—accepting that there is as yet no reliable biological indicator of tumour behaviour—the question of major surgery in the elderly man requires careful consideration. Presently there are few centres in Europe where this procedure would be considered in a man over 70 years and those between 65 and 70 demand a most critical assessment of their ability to withstand both the operation itself and the extensive staging lymphadenectomy that preceeds radical removal of the gland. Overall, it seems likely that there will continue to be an increase in the number of these procedures performed in fit elderly men until such time as questions regarding tumour biopotential are resolved.

Lymph node disease

Involvement of pelvic lymph nodes is a common finding at operation in those patients initially thought to have confined disease.[104] Such patients naturally carry a worse prognosis than those with organ confined disease[105] but in those elderly patients in whom the extent of the disease is unknown progression may only be presumed by a rising PSA level in conjunction with the imperfect evidence contained from computerized tomography scans. Taking the size of the primary into account, it is likely that PSA levels greater than 20 ng almost certainly indicate nodal spread;[72] such assumptions are however largely academic as the older patient is almost certain to be offered hormonal manipulation whether the source of this tumour marker is from the primary or secondary node disease.

Unlike the paucity of nodal information which is available in the majority of elderly patients who are not subjected to surgical staging, a small group of men may present with massive lymphadenopathy which is commonly palpable and frequently leads to obstructive renal failure. In such cases skeletal changes may or may not be present and serum marker levels are often greatly elevated. Despite the apparent hopelessness of such patients with renal disturbances, remarkable resolution of disease has been observed to occur in the node only group following hormone manipulation. Under these circumstances aggressive treatment with nephrostomy as necessary may allow high quality of life for significant periods (70 per cent 5 year survival) before reactivation of hormone resistant disease occurs.[51]

Bony metastatic disease

As noted above it is unfortunately the case that, within the UK, more than 50 per cent of men newly diagnosed with prostate cancer present with a positive bone scan. Although rare cases without systemic symptoms may on occasion be rationally placed in a 'delayed treatment' group, the great majority will receive immediate hormone manipulation as the treatment of choice.

Orchidectomy Orchidectomy, either total or subcapsular, remains the gold standard of therapy for metastatic prostatic cancer. Rapid lowering of serum testosterone follows this manoeuvre which allows early relief of painful symptomatology. Despite considerable pharmaceutical pressure to prove otherwise, the treatment is also cheaper by far than administration of luteinizing hormone releasing hormone (LHRH) analogues by regular monthly depot injection.

LHRH analogues Introduced during the last 15 years, LHRH analogues act by down regulation of luteinizing hormone release by paradoxical desensitization of receptors in the pituitary. During the initial 10 days treatment LH (and thus testosterone) levels actually rise ('flare' phenomenon) and may give rise to considerable pain with risk of skeletal complications particularly if the spine is severely affected by disease. Such complications due to flare may be avoided by pretreatment with antiandrogens such as cyproterone acetate which should be maintained for 1 week after the LHRH injection has been administered. However, most patients are not affected by these problems and LHRH agonists, which have been shown to be as effective as orchidectomy for metastatic disease, are widely prescribed as first-line therapy for patients who prefer not to consider surgical treatment.[64]

Total androgen ablation Ten years ago Labrie and associates reported that intraprostatic dihydrotestosterone levels remained high (40 per cent of precastrate levels) despite hormonal manipulation, and suggested far superior results could be obtained by combining an antiandrogen with the LHRH therapy (total androgen suppression).[106] These spectacular results were greeted with cynicism but a number of trials were subsequently commenced to test the hypothesis. Initially results seemed at best equivocal as regards quality of life[107] but recently the European Organization for Research and Treatment of Cancer (EORTC) study group have reported in abstract form that significant advantage may indeed be observed in the combination therapy group. The question now devolves—especially in the elderly population—to the problem of which particular subgroup of patients with advanced carcinoma should best benefit from this double therapy which carries significant financial implications.

Side-effects of systemic androgen deprivation   Regardless of the mechanism by which hormone manipulation is achieved, the reduction in serum testosterone leads to side-effects which some patients find troublesome. Initially hot flushes may be severe and although after a period of time (months) symptoms regress, additional treatment with progestational agents such as cyproterone acetate may be necessary. Loss of libido and reduction in penile size may cause anxiety and psychological disturbance even in elderly patients, and all such side-effects will require full discussion with both patient and spouse before treatment commences. For these reasons the search for pure antiandrogenic therapy has intensified in recent years.

Antiandrogens   Cyproterone acetate, the first antiandrogen to be described and used for many years in the treatment of sexual overactivity, is steroidal in nature and exhibits potent progestin-like activity. Thus effects at the target organ are reinforced by reduction in serum androgens. However, side-effects of fluid retention and thrombosis thought to be related to the steroidal configuration[108] may often preclude its use in elderly men with compromised cardiovascular systems. Libido is severely depressed and reduction in erectile potency with gynaecomastia frequently experienced. Nevertheless, by virtue of progestational properties, the drug can be helpful in reducing the hot flushes caused by other therapeutic agents. The agent is also frequently used in the UK after the failure of first-line therapy, often with surprisingly good short-term results.

Pure antiandrogens   The concept of a 'pure' antiandrogen which may block peripheral action of androgens whilst leaving circulating levels unaffected has yet to be attained. Central antagonist effects on the hypothalamic – pituitary negative feedback mechanism determine that LH and thus serum testosterone rises, so the dose of antiandrogen must be high enough to ensure that the elevated serum androgens are not causing breakthrough stimulation of cancer cells. Flutamide (Drogenil, Schering-Plough) is now licenced for the treatment of prostate cancer whilst ICI 176, 334 'Casodex' is undergoing multinational clinical phase III trials at the present time.

Flutamide, being non-steroidal, is not associated with either oedema or thromboembolism and does not suppress testicular function or libido.[109,110] Nausea, diarrhoea, and dizziness have however been described in a number of patients and may necessitate withdrawal of therapy.[111] In common with all antiandrogens, gynaecomastia (usually of mild degree) is reported in a majority of patients and the short half-life of the drug determines that three times daily dosage is necessary to maintain optimum blood levels of antiandrogen. ICI 176,334 'Casodex' has a half-life of 5–7 days and can thus be administered daily.[112] Few side-effects (apart from gynaecomastia) seem to be emerging with this drug and the results of the phase III monotherapy trials are awaited. Future developments may include the combination of antiandrogen with LHRH agonist therapy in view of recent EORTC data reported (see above) from total androgen suppression studies.

Treatment of the relapsed patient

There are no accepted treatment guidelines for those patients who have failed initial hormone manipulation, and at this point in the patient's illness it is becoming increasingly obvious to all concerned that quality of life is the overriding principle by which treatment should be judged. Scientific appraisal of second-line therapy is itself hindered by the differing definitions of what constitutes primary treatment failure; some will treat when the patient's symptom scores deteriorate whilst others initiate further therapy when biochemical markers such as PSA rise again; this usually occurs a significant interval before clinical evidence of treatment failure is apparent. Orchidectomy may be considered for second-line therapy on failure of initial medical treatment but the results are at best unpredictable and usually short lasting.[113]

Treatment with oestrogenic compounds is usually contraindicated in view of potential cardiovascular morbidity whilst reduction in adrenal androgens by medical or surgical ablation is rarely considered owing to the inherent risks of such measures in an elderly group of men who subsequently require gluco- and mineralocorticoid replacement. As mentioned above, treatment with cyproterone acetate is probably the commonest second-line therapy used in the UK despite equivocal evidence of efficacy in the relapsed patient.[114]

Recent studies on the mechanism of bone resorption in patients with advanced disease has demonstrated significant osteoclastled osteolysis both in normal and metastatic bone[115] but efforts to inhibit this process by disphosphonate therapy, whilst demonstrating a stabilizing effect on skeletal metabolism, have failed to show consistant clinical benefit in terms of pain control or time to progression.[116] It seems likely that diphosphonate therapy is unable to suppress the direct tumour stimulated osteoclast mediated changes seen in patients with advanced disease; results of ongoing studies are awaited.

Chemotherapy   Despite a number of clinical trials conducted chiefly in the USA and Europe in the last few years, evidence for the efficacy of systemic chemotherapeutic regimes remains tenuous. In general urologists in the UK have been very reluctant to subject their elderly patients to regimes with significant side-effects. Combination agents such as oestradiol and nitrogen mustard (Estracyt- Kabi Pharmacia) have been widely used and shown to be effective in some series. Suramin, an antihelmenthic agent, has been studied recently but results remain open to question; additionally the drug may have severe toxic side-effects, particularly on the spinal cord, and close surveillance during therapy is essential.

Radiation treatment   Localized 'single shot' radiation therapy to painful areas shown to be related to metastatic deposits remains the most useful single therapy available for treatment-relapsed patients. Pain relief is usually excellent and swift although further treatments cannot usually be given to the same area. Hemibody irradiation[117] may on occasion be beneficial but demands close cooperation between radiotherapist and urologist in view of likely problems related to marrow suppression. Strontium 89 has recently been introduced but experience is limited not least because of the very high cost of this agent.

Cancer care support

The unpredictable course of patients with advanced prostate cancer determines that such patients fit uneasily into the busy regular schedule of the clinical urologist. Under these circumstances the support available at specialist clinics by a dedicated nurse with extensive pain control experience is invaluable and offers patients the considerable placebo boost of intensive aftercare supplanted by medical intervention as necessary. Finally, nursing liaison between patient, doctor, and relatives ensures that quality of aftercare is maintained during the inevitable transition period between hospital and hospice care.

Key points

- Prostate cancer is now the commonest malignancy in elderly US males.
- Histological evidence of malignancy is found in 10–15 per cent of men undergoing resection of clinically benign tumors.
- The optimal management of localized prostate cancer by radical surgery, radiotherapy, or expectant treatment has yet to be determined.
- In patients with metastatic prostate carcinoma, 50 per cent will be dead within 2 years. Better performance status predicts for longer survival.
- Relief of outlet obstruction is the first requirement in a patient with prostate cancer.
- Orchidectomy remains the standard treatment for metastatic prostatic carcinoma.

• Support from dedicated nurses at specialized units plays an important role in the management of elderly patients.

References

1. Parkin, D.M., Stjernswärd, J., and Muir, C.S. (1984). *Bulletin of the World Health Organization*, **62**, 163–82.
2. Dhom, G. (1983). *Journal of Cancer Research Clinical Oncology*, **106**, 210–18.
3. Office of Population Censuses and Surveys (1989). *Monitor DH2-89*, (Cancer Research Campaign Factsheet 3.1. HMSO, London.
4. Registrar General (1975). *The registrar general's statistical review of England and Wales 1968–1970*. HM. Stationary Office, London.
5. Enstrom, J.E. and Austin, D.F. (1977). *Science*, **195**, 847–51.
6. Jewitt, H.J. (1984). *Journal of Urology*, **131**, 845–9.
7. Silverberg, E. and Holleb, A.I. (1975). *Cancer Journal for Clinicians*, **25**, 2–8.
8. Cantrell, B.B., Deklerk, D.P., Eggleston, J.C., Boitnott, J.K., and Walsh, P.C. (1981). *Journal of Urology*, **125**, 516–20. Silverberg, E. and Lubera, J.A. (1988). *Cancer Statistics CA*, **38**, 14–15.
9. Silverberg, E. and Lubera, J.A. (1989). *Cancer Statistics CA*, **39**, 3–20.
10. National Cancer Institute (1988). *Division of cancer prevention and control 1987 annual cancer statistics review*, NIH Publication No. 88, p. 2789. National Cancer Institute, Bethesda, MD.
11. Skeet, R.G. (1976). In *Scientific foundations of urology*, Vol. II, (ed. D.I. Williams and G.D. Chisholm, pp. 199–211. Heinemann.
12. Waterhouse, J.A., Hm Muir, C.S., Shanmugaratnam, K., and Powell, J. (1982). IARC Scientific Publication No. 15. IARC, Lyons.
13. Wynder, E.L., Mabuchi, K., and Whitmore, W.F. (1971). *Cancer*, **28**, 344–80.
14. Carter, H.B. and Coffey, D.S. (1990). *Prostate*, **16**, 39–48.
15. Dunn, J.E. (1975). *Cancer Research*, **35**, 3240–5.
16. Metlin, C. and Natarajan, N. (1983). *Prostate*, **4**, 223–331.
17. Murphy, G.P., Natarajan, N., Pontes, J.E., Schmitz, R.L., Smart, C.R., Schmidt, J.D., and Mettlin, C. (1982). *Journal of Urology*, **127**, 928–34.
18. Jackson, M.A., Ahluwalia, B.S., and Herson, J. (1977). *Cancer Treatment Reports*, **61**, 167–72.
19. Huben, R., Mettlin, C., Natarajan, N., Smart, C.R., Pontes, E., and Murphy, G.P. (1982). *Urology*, **20**, 585–8.
20. Feldman, A.R., Kessler, L., Myers, M.H., and Naughton, M.D. (1986). *New England Journal of Medicine*, **315**, 1394–7.

21. Cook, P.J., Doll, R., and Feelingham, S.A. (1969). *International Journal of Cancer*, **4**, 93–112.
22. US Bureau of Census (1986). *Statistical abstract of the United States 1987*, (107th edn.). US Government Printing Office, Washington DC.
23. Franks, L.M. (1956). *Lancet*, **II**, 1037–9.
24. Hirst, A.E. and Bergman, R.T. (1954). *Cancer*, **7**, 136–41.
25. Scott, R., Mutchnik, D.L., Laskowski, T.Z., and Schmalhorst, W.R. (1969). *Journal of Urology*, **101**, 602–7.
26. Holund, H. (1980). *Scandinavian Journal of Urology and Nephrology*, **14**, 29–35.
27. Clifton, J.A., Philipp, R.J., Loduvic, B.S., and Fowler, W.M. (1952). *American Journal of Medical Science*, **224**, 121–30.
28. Lundberg, S. and Berge, T. (170). *Scandinavian Journal of Urology and Nephrology*, **4**, 93–7.
29. Halpert, B., Sheehan, E.E., Schmarlhurst, W.R., and Scott, R. (1963). *Cancer*, **16**, 737–42.
30. Breslow, N., Chan, C.E., Dhom, G., et al. (1977). *International Journal of Cancer*, **20**, 680–8.
31. Guileyardo, J.M., Johnson, W.D., Walsh, R.A., Akazaki, K., and Correa, P. (1980). *Journal of the National Cancer Research Institute*, **65**, 311–16.
32. Yatani, R., Chigusa, K., Akazaki, K., and Stemmermann, G.N. (1982). *International Journal of Cancer*, **29**, 611–16.
33. Potosky, A.L., Kessler, L., Gridley, G., Brown, C.C., and Horm, J.W. (1990). *Journal of the National Cancer Institute*, **82**, 1624–8.
34. Beynon, L., Busuttil, A., Newsham, J.E., and Chisholm, G.D. (1983). *British Journal of Urology*, **55**, 733–6.
35. Silverberg, E. (1987). *Cancer*, **60**, 692–717.
36. Haenszel, W. and Kurihara, M. (1968). *Journal of the National Cancer Institute*, **40**, 43–68.
37. Meikle, A.W., Smith, J.A., and West, D.W. (1985). *Prostate*, **6**, 121–8.
38. Carroll, K.K. and Kohr, H.T. (1975). *Progress in Biochemical Pharmacology*, **10**, 308–53.
39. Whittemore, A.S., Paffenbarger, R.S., Anderson, K., and Lee, J.E. (1984). *Journal of Urology*, **132**, 1256–61.
40. Ross, R.K., Deapen, D., Cassagrande, J., Paganini-Hill, A., and Henderson, B.E. (1981). *British Journal of Cancer*, **43**, 233–5.
41. George, N.J.R. (1988). *Lancet*, **I**, 494–7.
42. Adolfsson, J., Rönström, J., Carstensen, T., Lowhagen, T., and Hedlund, P.O. (1990). *British Journal of Urology*, **65**, 611–14.
43. Johansson, J.E., Anderson, S.E., Krusemo, U.B., Adami, H-O., Bergstrom, R., and Kraaz, W. (1989). *Lancet*, **I**, 799–803.

44. Kramer, S.A., Spahr, J., Brendler, C.B., Glenn, J.F., and Paulson, D.F. (1980). *Journal of Urology*, **124**, 223–5.
45. Stamey, T.A., McNeal, J.E., Freiha, F.S., and Redwine, E. (1988). *Journal of Urology*, **139**, 1235–41.
46. Benson, R.C., Tomera, K.M., Zincke, H., Fleming, T.R., and Utz, D.C. (1984). *Journal of Urology*, **131**, 1103–6.
47. Goodman, C.M., Busuttil, A., and Chisholm, G.D. (1988). *British Journal of Urology*, **62**, 576–80.
48. McNeal, J.E., Bostwick, D.G., Kinrachuk, R.A., Redwine, E.A., Freiha, F.S., and Stamey, T.A. (1986). *Lancet*, **I**, 60–3.
49. Ewing, J. (1928). *Neoplastic diseases : a text book on tumors*. Saunders, Philadelphia, PA.
50. Sheldon, C.A., Williams, R.D., and Fraley, E.E. (1980). *Journal of Urology*, **124**, 626–31.
51. Sandhu, D.P.S., Mayor, P.E., Sambrook, P., and George, N.J.R. (1990). *British Journal of Urology*, **66**, 415–19.
52. Epstein, J.L., Paull, G., Eggleston, J.C., and Walsh, P.C. (1986). *Journal of Urology*, **136**, 837–9.
53. Reference not submitted.
54. Epstein, J.I. and Steinberg, G.D. (1990). *Cancer*, **66**, 1927–32.
55. Franzen, S., Giertz, G., and Zajicek, J. (1960). *British Journal of Urology*, **32**, 193–6.
56. Jönsson, G. (1971). *Scandinavian Journal of Urology and Nephrology*, **5**, 97–102.
57. Trasti, H., Nilsson, S., and Peterson, L.-E. (1979). *British Journal of Urology*, **51**, 135–9.
58. Smith, J.A. and Cho, Y.-H. (1990). *Urological Clinics of North America*, **17**, 769–77.
59. Jewitt, H.J. (1975). *Urological Clinics of North America*, **2**, 105–15.
60. Blute, M.L., Zincke, H., and Farrow, G.M. (1986). *Journal of Urology*, **136**, 840–3.
61. McMillen, S.M. and Wettlaufer, J.N. (1976). *Journal of Urology*, **116**, 759–60.
62. Parfitt, H.E., Smith, J.A., Gleidman, J.B., and Middleton, R.G. (1983). *Cancer*, **51**, 2346–50.
63. Hayashi, T., Taki, Y., Ikai, K., Hiura, M., Kiriyama, T., and Shizuki, K. (1987). *Prostate*, **10**, 275–9.
64. Kaisery, A., Tyrell, C.J., Peeling, W.B., and Griffiths, K. (1991). *British Journal of Urology*, **67**, 502–8.
65. Carter, H.B. and Isaacs, J.T. (1988). *Seminars in Urology*, **6**, 262–8.
66. Isaacs, J.T. and Coffey, D.S. (1981). *Cancer Research*, **41**, 5070–6.
67. de Voogt, H.J., Sucin, S., Sylvester, M., *et al.* (1990). In *EORTC genitourinary monograph*: prostate and testicular cancer, (ed. D.W.W. Newling and W.G. Jones), pp. 69–72. Wiley-Liss, New York.

68. Chodak, G.W., Keller, P., and Schoneberge, H.W. (1989). *Journal of Urology*, **141**, 1136–8.
69. Lee, F., Littrup, P.J., Torp-Pederson, S.T., Mettlin, C., Mehugh, T.A., Gray, J.M., Kumasaca, G.H., and McLear, R.D. (1988). *Radiology*, **168**, 389–94.
70. Optenberg, S.A. and Thompson, I.M. (1990). *Urological Clinics of North America*, **17**, 719–37.
71. Cooner, W.H., Mosley, B.R., Rutherfored, C.L., Jr., Beard, J.H., Pond, H.S., Terry, W.J., Igel, T.C., and Kidd, D.D. (1990). *Journal of Urology*, **143**, 1146–54.
72. Pantilides, M.L., Bowman, S.P., and George, N.J.R. (1992). *British Journal of Urology*, **70**, 299–303.
73. Wang, M.C., Valenzuela, Murphy, G.P., and Chu, T.M. (1979). *Investigative Urology*, **17**, 159–63.
74. Stamey, T.A., Yang, N., Hay, A.R., McNeal, J.E., Freiha, F.S., and Redwine, E. (1987). *England Journal of Medicine*, **317**, 909–16.
75. Lee, F., Littrup, P.J., Loft-Christensen, L., Kelly, B.S. JN, McHugh, T.A., Siders, D.S., Mitchell, A.E., and Newby, J.E. (1992). *Cancer*, **70**, 211–20.
76. Cooper, E.H., Armitage, T.G., Robinson, M.R.G., (Newling, D.W., Richards, B.R., Smith, P.H., Denis, L., and Sylvester, R. (1990). *Cancer*, **66**, 1025–8.
77. Fidler, I. (1984). *Cancer Treatment Reports*, **68**, 193–8.
78. Jonasson, O., Long, L., Roberts, S., McGrew, E., and McDonaid, J.H. (1961). *Journal of Urology*, **85**, 1–12.
79. Kuban, D.A., El-Mahdi, A.M., Schellhammer, P.F., and Bebb, T.J. (1985). *Cancer*, **56**, 961–4.
80. Paulson, D.F. and Cox, E.B. (1987). *Journal of Urology*, **138**, 90–1.
81. D'Amore, P.A. (1986). *Progress in Clinical and Biological Research*, **212**, 269–86.
82. Lowe, F.C. and Isaacs, J.T. (1984). *Cancer Research*, **44**, 744–52.
83. Nicolson, G.L. (1982). *Biochemica et Biophysica Acta*, **695**, 113–28.
84. Edelman, G. (1983). *Science*, **219**, 450–7.
85. Carter, H.B., Partin, A.W., and Coffey, D.S. (1989). *Journal of Urology*, **142**, 1338–41.
86. Foster, C.S., McLoughlin, J., Bashir, I., and Abel, P.D. (1992). *Human Pathology*, **23**, 381–94.
87. McLoughlin, J., Foster, C.S., Braum, P., et al. (1992). *Journal of Urology*, (In press).
88. Keer, H.N., Kozlowski, J.M., Tsai, Y.C., Lee, C., McEwen, R.N., and Grayhack, J.T. (1990). *Journal of Urology*, **143**, 381–5.
89. Blute, M.L., Nativ, O., Zincke, H., Farrow, G.M., Therneau, T., and Lieber, M.M. (1989). *Journal of Urology*, **142**, 1262–5.

90. Lee, S.E., Currin, S.M., Paulson, D.F., and Walther, P.J. (1988). *Journal of Urology*, **140**, 769–74.
91. McIntyre, T.L., Murphy, W.M., Coon, J.S., Chandler, R.W., Schwartz, D., Conway, S., and Weinstein, R.S. (1988). *American Journal of Clinical Pathology*, **8**, 370–3.
92. Ring, K.S., Karp, F.S., Olsson, C.A., O'Toole, K., Bixon, R., and Benson, M.C. (1990). *Prostate*, **17**, 155–64.
93. Partin, A.W., Isaacs, J.T., Treiger, B., and Coffey, D.S. (1988). *Cancer Research*, **48**, 6050–3.
94. Treiger, B. and Isaacs, J.T. (1988). *Journal of Urology*, **140**, 1580–6.
95. Manishen, W.J., Sivananthan, K., and Orr, F.W. (1986). *American Journal of Pathology*, **123**, 39–45.
96. Mundy, G.R., DeMartino, S., and Rowe, D.W. (1981). *Journal of Clinical Investigation*, **68**, 1102–5.
97. Magro, C., Orr, F.W., Manishen, W.J., Sivananthan, K., and Mokashi, S. (1985). *Journal of the National Cancer Institute*, **74**, 829–38.
98. Chackal-Roy, M., Niemeyer, C., Moore, M., and Zetter, B.R. (1989). *Journal of Clinical Investigation*, **84**, 43–50.
99. Strobel, E.S., Strobel, H.G., Bross, K.J., Winterhalter, B., Fiebig, H.H., Schildge, J.L., and Lohr, G.W. (1989). *Cancer Research*, **49**, 1001–7.
100. Budharaja, S.N. and Anderson, J.C. (1964). *British Journal of Urology*, **36**, 535–9.
101. Kabalin, J.N., Hodge, K.K., McNeal, J.E., Freiha, F.S., and Stamey, T.A. (1989). *Journal of Urology*, **142**, 326–31.
102. Scardino, P.T., Frankel, J.M., Wheeler, T.M. Meacham, R.B., Hoffman, S., Seale, C., Wilbanks, J.H., Easley, J., and Carlto, C.E., JN. (1986). *Journal of Urology*, **135**, 510–16.
103. Catalona, W.J. and Bigg, S.W. (1990). *Journal of Urology*, **143**, 538–44.
104. Zincke, H. (1990). *Seminars in Urology*, **8**, 175–83.
105. Steinberg, G.D., Epstein, J.L., Piantadosi, S., and Walsh, P.C. (1990). *Journal of Urology*, **144**, 1425–32.
106. Labrie, F., Dupont, A., Bélanger, A., et al. (1982). *Clinical Investigative Medicine*, **5**, 267–75.
107. Crawford, E.D., Eseinberger, M.A., McLeod, D.G., Spaulding J.T., Benson Dorr, F.A., Blumenstein, B.A., Davis, M.A., and Goodman, P.J. (1989). *New England Journal of Medicine*, **321**, 419–24.
108. Neumann, F. and Jacobi, G.H. (1982). *Clinical Oncology*, **1**, 41–66.
109. MacFarlane, J.R. and Tolley, D.A. (1985). *British Journal of Urology*, **57**, 172–4.
110. Neri, R.O. and Peets, E.A. (1975). *Journal of Steroid Biochemistry*, **6**, 815–19.
111. Furr, B.J.A. (1989). *Hormone Research*, **32**, 69–76.

112. Stone, A.R., Hargreave, T.B., and Chisholm, G.D. (1980). *British Journal of Urology*, **52**, 535–8.
113. Giuliani, L., Pescatore, D., Giberti, C., Martorana, G., and Natta, G. (1980). *European Urology*, **6**, 145–8.
114. Clarke, N.W., McClure, J., and George, N.J.R. (1992). *British Journal of Urology*, **69**, 64–70.
115. Clarke, N.W., Holbrook, I.B., McClure, J., and George, N.J.R. (1991). *British Journal of Cancer*, **63**, 420–3.
116. Rowland, C.G., Bullimore, J.A., Smith, J.B., *et al.* (1981). *British Journal of Urology*, **53**, 628–32.
117. Reference not submitted.

8

Bladder cancer

Michele Pavone-Macaluso and Vincenzo Serretta

Introduction

Recent data suggest that the mean human life expectancy at birth is 79.4 years for women and 73.8 years for men (Olshansky *et al.* 1990). The size of the problem of cancer in the elderly cannot be dismissed because of a presumed, unconfirmed concept of a less aggressive course of the disease. It has been calculated that by the year 2000, approximately 70 per cent of all cancers will occur in elderly patients (Monfardini and Chabner 1991). Bladder carcinoma has a peak incidence at 70 years and is relatively frequent in both elderly men and women.

The great majority of reports on bladder cancer make no special reference to the age of the patients. Brief mentions are made occasionally of age limits for radical surgery. Cystectomy is usually recommended, if indicated, for patients below 70 years of age. No account is provided for treatment of older patients, although they represent a considerable percentage of the overall number of affected patients.

As with the majority of cancers treatment options differ considerably according to the stage of the disease.

We consider four different groups of patients with the following conditions:

(1) carcinoma *in situ*;
(2) superficial papillary tumours;
(3) deeply infiltrating cancer;
(4) metastatic disease.

Carcinoma *in situ*

Carcinoma *in situ* is amenable to local treatment using either BCG or conventional chemotherapeutic agents. In the case of failure, cystourethrectomy is indicated. The issue of extensive surgery in the elderly is discussed below in the section devoted to deeply infiltrating cancer.

Local chemotherapy can be administered safely to old patients, inasmuch

Table 8.1 Superficial papillary bladder tumours: age groups and treatment modalities between 1980 and 1990 at the Department of Urology of Palermo

Age	Laser (1985–90)	TUR
< 30	–	6
31–40	–	23
41–50	7	34
51–60	16	200
61–70	22	284
	45	547
71–5	15	159
76–80	10	125
81–5	1	44
86–90	–	17
> 90	–	1
	26	346

as no systemic toxicity is to be expected with the currently available drugs. However, special care must be taken with the use of BCG, since severe and even lethal complications may follow systemic BCG absorption. Such severe consequences are relatively frequent after a difficult, traumatic catheterization (Steg *et al.* 1989). This is much more likely to occur in elderly men with prostatic enlargement, than in younger male patients.

Superficial papillary tumours

In general, no age limit can be established for the treatment of most cases of superficial bladder cancer, since transurethral resection (TUR), laser coagulation, and adjuvant intravesical treatment can be applied safely. Our experience is reported in Table 8.1, which shows that of 964 patients treated by laser or TUR for superficial bladder carcinoma, 372 (39 per cent) were older than 70 years. Anaesthesia for TUR does not usually represent a problem.

Deeply infiltrating cancer

Accepted forms of treatment are radical cystectomy and irradiation, with or without adjuvant or upfront chemotherapy. Only in rare cases is

Table 8.2 Patients, by age group, submitted to primary chemotherapy at the Department of Urology of Palermo between 1985 and 1990

Age	Number of patients
< 70	32 (50.8%)
71–5	21 (33.3%)
76–80	6 (9.6%)
81–5	4 (6.3%)

Chemotherapy regimen (q. 21 days) × 4 cisplatin 70 mg/mq day 1, methotrexate 40 mg/mq days 8 and 15.

partial cystectomy or extensive TUR indicated. Postoperative systemic chemotherapy is poorly tolerated and its real value is still controversial. It is therefore wiser to withhold it in old, relatively frail patients, even when they have been considered fit for the operative procedure. The tolerance of preoperative chemotherapy is much better, although its real value still requires to be established. No clear-cut overall survival and advantage has emerged so far, but patients showing a complete clinical response have a far better survival than non-responders. Even a partial response carries a relatively good prognosis, irrespective of the definitive treatment (radical cystectomy, irradiation, extensive TUR).

Pre-emptive chemotherapy *per se* is well tolerated, even in old patients. In our experience with preoperative cisplatin and methotrexate, neither the response rate nor dose intensity were different in the older age group compared with younger patients (Tables 8.2, 8.3 and 8.4). Therefore, although these data need further confirmation, first-line chemotherapy can be considered a legitimate option even for patients aged 75 or 80 years. If no response is obtained, the patient can probably be spared a cystectomy, as the prognosis, in non-responders to pre-emptive chemotherapy, is very poor. However, if a complete response is obtained, an extensive TUR

Table 8.3 Number of full cycles of chemotherapy by age group

Age	Number of cycles
< 70	3.3
71–5	2.9
76–80	3.6
81–5	3.5

Table 8.4 Objective response (complete and partial) to primary chemotherapy by age group

Age	Objective response (%)
< 70	45
71–5	47.6
76–80	50
81–5	33

(through the whole thickness of the bladder) followed by irradiation may be preferred to total cystectomy in the very old patient. As far as irradiation is concerned, present evidence shows that old age does not represent an absolute contraindication if high energy is employed (Hope-Stone 1986).

The risk of extensive surgery in patients older than 70 years is still a matter of discussion. A few reports of total cystectomy merely state that patients aged between 70 and 80 were included in the surgical series. The surgical risk clearly depends, among other factors, on the risk of general anaesthesia (Djokovic and Hedley-Whyte 1979). Hosking *et al.* (1989) reported an overall mortality rate of only 8.4 per cent among 795 surgical patients older than 90 years. Denney and Denson (1972), two decades ago, reported an overall mortality rate of 29 per cent in patients older than 90 who underwent similar surgical procedures. Although a direct comparison of these data cannot be made, the striking difference in survival suggests an important improvement in postoperative survival in elderly patients.

The surgical risk in urological procedures as such has been evaluated differently by various authors. Bollack *et al.* (1989) reported on a group of 32 patients, whose age ranged from 70 to 86 years with an average age of 74.6 years. The mortality rate was 3.1 per cent, which is similar to that reported in the literature. The authors concluded that radical cystoprostatectomy or pelvic anterior exenteration is feasible in elderly patients with a low mortality rate and an acceptable morbidity. Radical cystectomy produced a good quality of life in operated patients and this is of special interest, since the active life expectancy of patients between 70 and 80 years of age is 5–8 years.

An opposite view was taken by Studler and Haschek (1978), who reviewed 331 patients over 70 years of age with bladder cancer, of whom 193 (58 per cent) suffered from deeply infiltrating disease. Most patients were treated with TUR rather than cystectomy, mainly because of the consideration that arteriosclerotic and cerebral changes were frequently present making urinary diversion very difficult for these patients to tolerate. They recommended that the indications for surgery must be carefully evaluated for each patient and made the following statement:

Table 8.5 Radical cystectomies with simple urinary diversion procedures performed at the Urologic Clinic of Palermo between 1985 and 1990, by age group

Age	Number of cystectomies
40–50	1 (1.8%)
51–60	11 (20.4%)
61–70	28 (51.9%)
> 70	14 (25.9%)

although 70 years of age is a random choice, one should speak correctly of the biological age.

In our experience, the perioperative mortality of radical cystectomy adopting simple procedures such as a Bricker conduit for urinary diversion is not higher in elderly than in younger patients. Table 8.5 shows the number of radical cystectomies, without continent diversion, according to age groups, performed in our institute between 1985 and 1990. Fourteen patients (26 per cent) were older than 70 years and no perioperative deaths occurred among them. However, the operating time should be kept to a minimum and very complicated methods of urinary diversion should be avoided (Pavone-Macaluso 1990).

Metastatic disease

Systemic chemotherapy is indicated in patients with metastatic bladder cancer. The chemotherapeutic regimens in elderly patients, usually weakened by the systemic spread of the tumour, must take into account both the toxicity of the treatment and the quality of life of the patient.

However, in our experience the tolerance of systemic chemotherapy does not differ in patients with metastatic disease compared with patients with locally advanced tumours who have similar performance status (Ingargiola *et al.* 1988).

Conclusions

From our own experience and from a review of the pertinent literature it appears that, even for a formidable surgical procedure such as cystectomy and urinary diversion, an upper age limit cannot be established and that patients older than 70, if free from major physical illnesses, tolerate the impact of surgery surprisingly well. Thus in treatment of bladder cancer, as well as for other fields of surgery, experience and common sense are essential to select the most effective treatment for each individual patient.

Key points

- The peak age incidence of bladder carcinoma is 70 years.
- Transurethral resection under general anaesthesia or epidural block is the treatment of choice for superficial papillary carcinomas.
- Deeply infiltrative cancer may need first-line chemotherapy which is effective in all ages. Radical cystoprostatectomy or pelvic anterior exenteration can be carried out with low mortality rates in elderly patients.
- Tolerance to chemotherapy for metastatic disease is unrelated to age.

References

Bollack, C., Jacqmin, D, Cuveleir, G., and Bertrand, D. (1989). Radical cystectomy in patients over 70 years. In *Therapeutic progress in urological cancers*, (ed. G.P. Murphy and S. Khoury), pp. 613–16. A.R. Liss, New York.

Denney, J.L. and Denson, J.S. (1972). Risk of surgery in patiens over 90. *Geriatrics*, **27**, 115–18.

Djokovic, J.L. and Hedley-Whyte, J. (1979). Prediction of outcome of surgery and anaesthesia in patients over 80. *Journal of the American Medical Association*, **242**, 2301–6.

Hope-Stone, H.F. (1986). Urological malignancy. In *Radiotherapy in clinical practice*, (ed. H.F Hope-Stone), pp. 27–67. Butterworth, London.

Hosking, M.P., Warner, M.A., and Lobdell, C.M. (1989). Outcomes of surgery in patients 90 years of age and older. *Journal of the American Medical Association*, **261** (13), 1909–15.

Ingargiola, G.B., Lamartina, M., Cassata, G., Corselli, G., Serretta, V., Caramia, G., Rizzo, F.P., and Pavone-Macaluso, M. (1988). The cisplatin – methotrexate neo-adjuvant protocol as first-line therapy of bladder cancer. Present experience in Palermo. In *Management of advanced cancer of prostate and bladder*, (ed. P.H. Smith and M. Pavone-Macaluso), pp. 579–83. A.R. Liss, New York.

Monfardini, S. and Chabner, B. (1991). Joint NCI—EORTC consensus meeting on neoplasia in the elderly. *European Journal of Cancer*, **27**, 653–4.

Olshansky, S.J., Carnes, B.A., and Cassel, C. (1990). In search of Methuselah: estimating the upper limits of human longevity. *Science*, **250**, 634–40.

Pavone-Macaluso, M. (1990). Editorial note: organ reconstruction and conservation in bladder cancer. In *Urological oncology. Reconstructive surgery, organ conservation and restoration of function. Progress in clinical and biological research*, EORTC Genitourinary Group Monograph 10, (ed. P.H. Smith and M. Pavone-Macaluso), pp. 143–5. Wiley-Liss, New York.

Steg, A., Leleu, C., Debre' B., Boccon-Gibod, L., and Sicard, D. (1989).

Systemic bacillus Calmette-Guerin infection in patients treated by intravesical BCG therapy for superficial bladder cancer. In *BCG in superficial bladder cancer. Progress in clinical and biological research*. EORTC Genitourinary Group Monograph 6, (ed. F.M.J. Debruyne, L. Denis, and A.P.M van der Meijden), pp. 325–34. A.R. Liss, New York.

Studler, G., Haschek, H. (1978). Transurethral resection or cystectomy? An attempt to answer this question by an evaluation of 688 patients with carcinoma of the bladder (with special reference to age, sex, stage and the presence of associated disease). In *Bladder tumours and other topics in urological oncology*, (ed. M. Pavone-Macaluso, P.H. Smith, and F. Edsmyr), pp. 193–200. Plenum Press, New York.

9

Colorectal cancer

J.M.A. Northover

Incidence

Colorectal cancer newly afflicts 137 000 people each year in Europe, and of these 85 000 will die of the disease. The overall death rate for the disease has not changed for several decades, though the age-adjusted mortality has improved in all age groups. This apparent anomaly is accounted for by the gradual ageing of the population, so that more cases are occurring as the decades pass. Colorectal cancer is predominantly a disease of old age; the median age at presentation in the UK is 69 years. The disease is four times as common in the over 65s compared with the 45–64-year-old band (Baranovsky and Myers 1986).

Aetiology

Our understanding of the aetiology of colorectal cancer has developed considerably in the past decade. The changes occurring in bowel epithelial cells that lead to invasive malignancy are characterized in the concept of the adenoma—carcinoma sequence, in which normal cells become dysplastic and aggregate into adenomas, a few of which grow in size and eventually become malignant. The most likely series of events leading to bowel cancer includes the initiation of small adenomas by environmental (i.e. dietary) carcinogens, and the growth of some of these under the appropriate conditions to form large adenomas, and then cancer. These stages are encouraged by dietary fat, which increases the amount of bile acids and cholesterol in the bowel. These substances are acted upon by bacteria to produce cancer promoters. Some food constituents, particularly fibre, mollify the effects of the promoters, in the case of fibre by hurrying them out of the bowel. It has been known for a long time that diet and the environment is not the whole story; in particular, it was apparent many decades ago that in some people inherited gene defects might predispose them to bowel cancer. Of late, this has become a very important area of research, and has given insights into the way in which environmental agents might actually achieve their effects: it is almost certain that 'environmental' factors cause the same mutations in somatic cells that are inherited in 'cancer families'.

The increasing incidence of colorectal cancer with advancing age has been an area of active research in the past few years, and data that add weight to the foregoing explanation of colorectal carcinogenesis have come to light. Two Italian groups have shown that the cellular kinetic changes that are thought to occur prior to the development of dysplasia are more common in the 'normal' mucosa of the ageing rectum (Paganelli *et al.* 1990; Roncucci *et al.* 1988). A Dutch study has suggested that dietary fibre intake in elderly people is significantly lower than in younger age groups, and that this correlates negatively with total bile acid concentration in faeces (Nagengast *et al.* 1988). This in turn might explain the increasing prevalence of bowel cancer with age. Another possible explanation of the increasing prevalence with age of bowel cancer and other common tumours is immunesenescence, a process which is associated with thymic involution and perhaps failing immune surveillance (Kaesberg and Ershler 1989). However, bowel cancer, and indeed the other common solid tumours, are not amongst those malignancies seen with increased frequency in immune deficient states, including AIDS and in transplant subjects.

Natural history and clinical presentation

At least 50 per cent of cases have demonstrably incurable disease at the time of presentation. Prognosis differs enormously in elective presentation – almost all cases presenting at Dukes' stage A are curable, whilst stage C^2 disease carries a 5 year survival of only 20 per cent. The proportion presenting as emergencies, with obstruction or perforation, varies between 20 and 50 per cent in different areas (Braun 1986; Waldron *et al.* 1986). Emergency presentation is associated, stage for stage, with a worse prognosis. There have been rather intriguing reports that the larger proportion of right-sided cancers seen in young patients (thought to be related to genetic predisposition) may be seen also in the elderly, particularly in women (Butcher *et al.* 1985; Jass 1991).

The elderly tend to present with more advanced disease (Bader 1986), though this may not be true world-wide (Adloff *et al.* 1986; Irvin 1988). The proportion who reach hospital only after the development of a life-threatening complication is 25 per cent above the population average, and the perioperative mortality in this high risk group is markedly above that seen in younger patients (Waldron *et al.* 1986).

With these principles in mind, it is worth reviewing recent studies which have examined surgical results in elderly patients. Several groups have reported non-comparative audits of results of radical and/or non-radical surgery. The most important of these emanates from the UK Large Bowel Cancer Project, in which many British surgeons have allowed unselected collation of the results of their surgical experience in this field (Fielding *et al*. 1989). More than three-quarters of post-operative deaths occurred in the 46 per cent of the study population who were over age 70 years old; more than half of the deaths were due to cardiopulmonary complications. In hospital mortality was four times greater in the above 70s compared with younger patients. Fielding and his colleagues concluded that results might be improved by earlier diagnosis to try to avoid emergency presentation, better post-operative monitoring, and a wider use of local treatments. In a series of 84 octogenarians, Wagner *et al*. (1987) found that 55 per cent had intercurrent disease, and that 27 per cent presented with obstruction or perforation. Their main conclusion was that every effort should be made to avoid emergency presentation, with its consequent increase in morbidity and mortality—as the authors put it 'It would seem that diagnostic and therapeutic nihilism in the elderly is inappropriate.'

Management

Surgery has been the mainstay of treatment of colorectal cancer. Much research effort has been expended on trying to define the role, if any, for adjuvant radiotherapy and chemotherapy, but there remains much uncertainty. Non-surgical primary therapy, using radiation, lasers, and electrocoagulation, has a small part to play, but in the elderly may offer approaches that are less life-threatening than surgery in appropriate cases.

Radical surgery

The range of surgical procedures available for the management of colorectal cancer is large, particularly in rectal cancer. Whereas in the early decades of this century rectal excision and permanent colostomy was the only radical treatment available for rectal cancer, today many patients are treated by sphincter-conserving resection, while a few undergo non-radical local excision in hope of cure.

There is much discussion of the relative merits of different surgical approaches in the elderly patient. Should sphincter-conserving surgery be used more or less often in this group, given the increased risk of post-operative complications as the price of colostomy avoidance? Is quality of life enhanced more by avoiding post-operative complications or by preventing colostomy? Even if the aged sphincters are saved, can they cope with the demons imposed by the soft or liquid effluent that they must hold back? (Christiansen 1988). If they cannot cope well, is the elderly patient better off with a colostomy anyway? Does the increased risk of local recurrence after local excision (with the need for further local excision or even subsequent radical surgery) offset the advantages of diminished morbidity and mortality? All these questions are matters of contemporary debate, and highlight the move towards addressing quality of life questions recently, particularly as improvements in cure rate seem so hard to achieve.

Yet surgeons must rely mainly on their instinct, 'clinical acumen', and in particular their view of 'elderliness', to deal with these questions in their day-to-day practice. There are few studies which address these questions scientifically, perhaps because it has proved so difficult to apply scientific method to quality of life issues. If the 'quality of life year (QuaLY)' were an easily defined and reliable tool, it would find ample use in this difficult area. Most studies are retrospective and seek to compare the outcome of a particular procedure in patients above and below a chosen age, often 70 or 75 years. Such studies are open to all sorts of bias, however well and ardently the authors recognize and discuss or try to minimize the effect of these potential pitfalls. The general result is that authors come to the rather vague conclusion that age *per se* should not be a bar to whatever surgical approach seems appropriate on technical grounds (Bader 1986; Lewis and Khoury 1988; De-Martino *et al.* 1991). This is of little help in the management in the individual case.

Non-radical therapy

Non-radical measures have their place, albeit a rather limited one, in attempts at cure in a few patients and in a larger proportion in whom only palliation is sought. In elderly patients these approaches have distinct advantages if the general state militates against surgery. Intracavitary irradiation for cure—using either beam therapy or implantation – electrocoagulation, and latterly lasers have their advocates, while all these modalities have a potentially wider role in palliation (Cutsem, Boonen et al. 1989).

The thrust of surgical development in rectal cancer has been towards less mutilating surgery without compromising the chance of cure. While this has led primarily to radical sphincter-saving surgery, another important strand has been the development of non-radical procedures, for both cure and palliation. Transanal excision of rectal tumours has been used in a small proportion of patients, around 5 per cent in published series. This avoids the morbidity and mortality of laparotomy, while offering a good prospect of cure in small tumours confined to the bowel wall and with better than poor differentiation (Balslev *et al.* 1986; Gall and Hermanek 1988; Snyder

1985; Whiteway *et al.* 1985). Others have described different techniques, most notably endoscopic transanal resection (ETAR), in which a urological resectoscope is used with good results (Berry *et al.* 1990). The drawback of this method is that a coherent surgical specimen is not available for the pathologist to examine for evidence of adequate excision.

Alternatives and adjuncts to surgery

While it is generally accepted that surgery has the primary role in treatment, for the elderly frail patient radiotherapy has been shown by careful clinical study to offer a good alternative in low rectal cancer (Papillon *et al.* 1989); treatment failure may occur in only around 15 per cent, the great majority of patients avoiding a colostomy. While chemotherapeutic agents are usually avoided in the elderly, more research is needed in this age group to define better their roles and contraindications (Mulder 1990; Neilan 1985).

While the role of adjuvant therapy in colorectal cancer remains far from clear, it appears that clinicians are less likely to use radiotherapy or chemotherapy in their older patients (Mor *et al.* 1985). It is not clear whether this is because of a perception of increased risk of side-effects, thereby diminishing quality of life, or whether there is seen to be a greater need to be 'sure' that everything possible has been done to ensure cure in the younger patient.

Screening

Although in the USA many clinicians accept the American Cancer Society's advice that regular screening is 'a good thing' in middle and old age, their colleagues in other parts of the world are less convinced. There is a dearth of data on the utility of screening in the elderly, in whom all the common cancers are most prevalent (Stults 1986), and in whom premalignant colorectal adenomas are so common (DiSario *et al.* 1991). In their attempts to elucidate this question, clinical researchers in Europe exclude from their randomized trials many at risk for colorectal cancer, i.e. those beyond the age of 70. The main arguments for this exclusion are that people beyond this age may not live to see the benefit, if any, of early diagnosis, and that poor compliance in the elderly would dilute the trial result. There is also, perhaps, the judgement that in a disease which predominantly affects the elderly, 'You have to die of something, sometime.' There is also the cost/benefit argument on behalf of society, which would have us believe that mass screening, if shown to be effective, should be targeted at the productive portion of the populace, rather than the elderly net consumers, prolongation of whose lives is merely a sap on the welfare state.

Neither of these arguments, recently described as 'Orwellian', should be acceptable in a civilized society (Fielding *et al.* 1989). Though death must come to us all, the often painful death in bowel cancer is one that should be avoided if possible. Moreover, if screening were to decrease the excess proportion of the elderly who present *in extremis* with obstruction or perforation, this would not only save them the enormous distress of this experience and probably give them a proportionately greater chance of cure from the disease, it would also diminish the cost to society of complex care in the emergency setting. Whether any putative extension of life due to screening in the elderly would be seen by them to be worth the 'price' of the discomfort of cancer treatment is an open question. In Europe active promotion of bowel cancer screening in the middle aged or elderly is discouraged until efficacy is demonstrated by randomized trials. In the USA, however, the reverse obtains—screening is encouraged, by the American Cancer Society and others, until efficacy is disproven (Robie 1989), only 20 per cent of Maryland practitioners believing that the elderly do not need regular sigmoidoscopy (Weisman *et al.* 1989). If the sweeping generalization that quality of life is more important than quantity in the elderly holds any truth (Robie 1989), surely the American tendency to assume the efficacy of screening and preventive measures until proven otherwise is overly invasive. Is it really acceptable not only to submit the elderly to regular surveillance but also to try to persuade them to change their diet and cooking methods substantially? (Bongiorno 1988).

Summary

Bowel cancer is common in all Western countries, and is becoming more prevalent in many others. It is remarkable that, although 50 per cent of patients with this disease are aged 70 and above, most research effort has been directed at younger patients, often with the specific exclusion of the elderly in randomized trials. With little extra imagination, therapeutic benefit in the elderly could be assessed very usefully in the fields of surgical strategy, adjuvant therapy, and screening. Only by such effort will we discover more about ways to defer death appropriately and to improve quality of life in this large segment of the bowel cancer population.

Key points

- The median age at presentation of colorectal cancer patients is 69 years.

- Dysplastic changes are more frequent in the mucosa of the ageing rectum and dietary fibre intake is significantly lower in older age groups.
- Quality of life after colostomy and sphincter-saving operations have not been studied in the elderly.
- Radiotherapy can offer a good alternative to surgery for frail patients with low rectal carcinoma.
- The role of adjuvant therapy in the elderly is not defined.
- Screening by faecal occult blood/sigmoidoscopy may be of value but its efficacy in all age groups remains unproven.

References

Adloff, M., Arnaud, J.P., Schloegel, M., Thibaud, D., and Bergamaschi, R. (1986). Colorectal cancer in patients under 40 years of age. *Dis. Colon Rectum*, **29** (5), 322–5.

Bader, T.F. (1986). Colorectal cancer in patients older than 75 years of age. *Dis. Colon Rectum*, **29**, 728–32.

Balslev, I., Pedersen, M., Teglbjaerg, P.S., Hanberg-Sorensen, F., Bone, J., Jacobsen, N., Overgaard, J., Sell, A., Bertelsen, K., and Hage, E. (1986). Major or local surgery for cure in early rectal and sigmoid carcinoma—a prospective evaluation. *Eur. J. Surg. Oncol.*, **12** (4), 373–7.

Baranovsky, A. and Myers, M.H. (1986). Cancer incidence and survival in patients 65 years of age and older. *Ca*, **36** (1), 26–41.

Berry, A.R., Souter, R.G., Campbell, W.B., Mortensen, N.J., and Kettlewell, M.G. (1990). Endoscopic transanal resection of rectal tumours—a preliminary report of its use. *Br. J. Surg.*, **77** (2), 134–7.

Bongiorno, C.P. (1988). Appropriate prevention and detection of gastrointestinal neoplasma in the elderly. *Clin. Geriatr. Med.*, **4** (1), 222–33.

Braun, L. (1986). Prognosis of colorectal cancer in patients over 80. *Dtsch. Med. Wochenschr.*, **111** (49), 1869–73.

Butcher, D., Hassanein, K., Dudgeon, M., Rhodes, J., and Hames, F.F. (1985). Female gender is a major determinant of changing subsite distribution of colorectal cancer. *Cancer*, **56** (3), 714–16.

Christiansen, J. (1988). Place of abdominoperineal excision in rectal cancer. *J.R. Soc. Med.*, **81** (3), 143–5.

De-Martino, D., Joly, A., Assenza, M., Arnaud, A., Sastre, B., and Sarles, J.C. (1991). Cancer of the rectum in elderly patients over the age of seventy-five years. Results of the surgical treatment. *Ann. Chir.*, **45** (3), 218–21.

DiSario, J.A., Foutch, P.G., Mai, H.D., Pardy, K., and Manne, R.K. (1991). Prevalence and malignant potential of colorectal polyps in asymptomatic, average-risk men. *Am. J. Gastroenterol.*, **86** (8), 941–5.

Fielding, L.P., Phillips, R.K., and Hittinger, R. (1989). Factors influencing mortality after curative resection for large bowel cancer in elderly patients. *Lancet*, **1** (8638), 595–7.

Gall, F.P. and Hermanek, P. (1988). Cancer of the rectum—local excision. *Surg. Clin. North Am.*, **68** (6), 1353–65.

Irvin, T.T. (1988). Prognosis of colorectal cancer in the elderly. *Br. J. Surg.*, **75** (5), 419–21.

Jass, J.R. (1991). Subsite distribution and incidence of colorectal cancer in New Zealand, 1974–1983. *Dis. Colon Rectum*, **34** (1), 56–9.

Kaesberg, P.R. and Ershler, W.B. (1989). The importance of immunesenescence in the incidence of malignant properties of cancer in hosts of advanced age. *J. Gerontol.*, **44** (6), 63–6.

Lewis, A.A.M. and Khoury, G.A. (1988). Resection for colorectal cancer in the very old: are the risks too high? *Br. Med. J.*, **296**, 459–61.

Mor, V., Masterson-Allen, S., Goldberg, R.J., Cummings, F.J., Glicksman, A.S., and Fretwell, M.D. (1985). Relationship between age at diagnosis and treatments received by cancer patients. *J. Am. Geriatr. Soc.*, **33** (9), 585–9.

Mulder, P. (1990). Drug treatment of cancer in elderly patients. *Tijdschr. Gerontol. Geriatr.*, **21** (4), 173–6.

Nagengast, F.M., van-der-Werf, S.D., Lamers, H.L., Hectors, M.P., Buys, W.C., and van Tongeren, J.M. (1988). Influence of age, intestinal transit time, and dietary composition on fecal bile acid profiles in healthy subjects. *Dig. Dis. Sci.*, **33** (6), 673–8.

Neilan, B.A. (1985). Management of cancer in the elderly. Implications of the aging process. *Postgrad. Med.*, **77** (8), 148–9.

Paganelli, G.M., Santucci, R., Biasco, G., Miglioli, M., and Barbara, L. (1990). Effect of sex and age on rectal cell renewal in humans. *Cancer Lett.*, **53** (2), 117–21.

Papillon, J., Montbarbon, J.F., Gerard, J.P., Chassard, J.L., and Ardiet, J.A. (1989). Interstitial curietherapy in the conservative treatment of anal and rectal cancers. *Int. J. Radiat. Oncol. Biol. Phys.*, **17** (6), 1161–9.

Robie, P.W. (1989). Cancer screening in the elderly. *J. Am. Geriatr. Soc.*, **37** (9), 888–93.

Roncucci, L., Ponz-de-Leon, M., Scalmati, A., Malagoli, G., Pratissoli, S., Perini, M., and Chahin, N.J. (1988). The influence of age on colonic epithelial proliferation. *Cancer*, **62** (11), 2373–7.

Snyder, S.K. (1985). Surgical treatment of low-lying carcinoma of the rectum. *Clin. Geriatr. Med.*, **1** (2), 485–92.

Stults, B.M. (1986). Preventive cancer care for the elderly. *Front. Radiat. Ther. Oncol.*, **20**, 182–91.

Van-Cutsem, E., Boonen, A., Geboes, K., Coremans, G., Hiele, M., Van-trappen, G., and Rutgeerts, P. (1989). Risk factors which determine the long term outcome of neodymium-YAG laser palliation of colorectal carcinoma. *Int. J. Colorectal Dis.*, **4** (1), 9–11.

Wagner, H.E., Aebi, U., and Barbier, P.A. (1987). Colorectal carcinoma in old age. *Schweiz. Med. Wochenschr.*, **117** (41), 1571–6.

Waldron, R.P., Donovan, I.A., Drumm, J., Mottram, S.N., and Tedman, S. (1986). Emergency presentation and mortality from colorectal cancer in the elderly. *Br. J. Surg.*, **73** (3), 214–16.

Weisman, C.S., Celentano, D.D., Teitelbaum, M.A., and Klassen, A.C. (1989). Cancer screening services for the elderly. *Public Health Rep.*, **104** (3), 209–214.

Whiteway, J., Nicholls, R.J., and Morson, B.C. (1985). The role of surgical local excision in the treatment of rectal cancer. *Br. J. Surg.*, **72** (9), 694–7.

10

Ovarian cancer

Matti S. Aapro

Introduction

This chapter will not review basic facts on ovarian cancer but will try to highlight particular aspects related to patients above the age of 65–70 years old, who constitute the majority of those with this disease. As in all chapters throughout this book, we emphasize the fact that the optimal approach for younger patients should be the approach used for the elderly patient with no objective, well-established contraindication for curative or palliative treatment.

Epidemiology

A recent publication by La Vecchia *et al.* (1992) discusses trends in cancer mortality in Europe and highlights the stability of ovarian cancer mortality in Northern Europe (the rate being between 9 and 10/100 000), while there is a rise in incidence in Southern Europe, with rates varying from 3 to 7/100 000, still much below those of the Nordic countries. This means that more than 20 000 European women (excluding the former Soviet Union) will die from ovarian cancer each year, compared with 13 000 in the USA (Boring *et al.* 1992). Of these 20 000 women, more than half are aged above 70, and the high incidence (rates from 60 to 100/100 000, Geneva Cancer Registry) of ovarian cancer in elderly women is not decreasing, although it has diminished in younger women (La Vecchia *et al.* 1992). As the European population as a whole is becoming older, one can expect more and more cases of ovarian cancer in the elderly in the years to come.

Pathology, clinical presentation, and outcome

Ovarian cancer in elderly women is almost exclusively of the epithelial type. While SEER (Surveillance, Epidemiology, and End-Results) data (Yancik *et al.* 1986) indicate that older patients present with more advanced disease, this may be biased by the inclusion of non-epithelial cancer in the younger age group. Lawton and Hacker (1990) have found a similar proportion of patients with stage III and IV ovarian cancer in those patients aged below 70 (35 out of 52) or above 70 years (21 out of 30). It may also be true that ovarian cancer, which can progress silently, presents at a more advanced stage in elderly patients, as do other gynaecological malignancies, because elderly women may not have regular gynaecological examinations (Holmes and Hearne 1981). However, a formal study of this question is lacking. The SEER and other data indicate a worse prognosis in older patients, possibly because of the more advanced stages of disease in the elderly population, because, when included in a multifactorial analysis, age is not an independent prognostic indicator for ovarian cancer (Redman *et al.* 1986). The data presented in this chapter indicate that age is not a major determinant of outcome, but have been derived from studies in which many elderly women were not included because of concomitant disease. This is a cofactor which has to be taken into account when looking at mortality in this particular population (Castiglione *et al.* 1990).

Surgical approach

Kennedy *et al.* (1989) reported that 22 of 114 patients aged above 75 presenting with gynaecological cancer at the Cleveland Clinic Foundation could not undergo surgery because of medical problems. Of the 13 patients operated on for ovarian cancer, four were optimally debulked, five suboptimally and four could not be debulked. These figures do not differ much from those in younger patients. However, they do not indicate the number of patients who were not referred to the centre for treatment. An idea of the selection bias is given by Samet *et al.* (1986) who found that while 75 per cent or more of the patients aged 65 or less diagnosed with ovarian cancer in New Mexico were operated on, only 36 per cent of those above age 65 had a surgical procedure. Kennedy *et al.* (1989), Lawton and Hacker (1990), and Krauer (1981) argue that careful attention to the nutritional status of patients submitted to laparotomies is essential, and that the higher post-operative mortality rate in elderly patients observed in some series, but not all, is heavily influenced by the patient's general condition, not the patient's age *per se*. Mortality in carefully prepared patients above 80 submitted to major gynaecological surgery may be as low as 3 per cent.

Radiotherapy

Presently this is a rarely used modality for ovarian cancer in both elderly and younger patients. Issues are discussed in this volume (Chapter 3). In 30 patients receiving various modalities of radiation therapy for gynaecological cancer, doses were reduced in ten instances according to Kennedy *et al*. (1989).

Chemotherapy

Any discussion of chemotherapy, its tolerance, and the results obtained in elderly people has to take into account the enormous selection bias of present day chemotherapy studies. While in the USA an age limit did not exist in the inclusion criteria of the Eastern Cooperative Oncology Group (ECOG) studies, Begg and Carbone (1983) have shown that many patients over 70 seen at ECOG institutions were not included in ovarian cancer trials. They reported that 10 per cent of ovarian cancer study patients were above the age of 70, compared with 17 per cent ovarian cancer patients above age 70 among those seen in ECOG centres. This already represents a referral selection, since 40 per cent or more of the American ovarian cancer patients are aged 70 or more. The European studies which we could review have usually had an age limit, and therefore data on really elderly patients are almost non-existent.

Tolerance of chemotherapy among ovarian cancer patients was first studied by Begg and Carbone (1983), in 630 patients aged less than 70 and in 72 patients aged 70 or more years, all of whom had entered ECOG studies. They reported that 26 per cent had significant haematological toxicity (platelets below 50 000/mm^3 or white blood cells below 2000/mm^3) among elderly people, compared with 14 per cent among younger patients (Table 10.1). Infections were also more frequent while vomiting was less frequent among elderly patients. Details are not given in their publication, and one has to assume that the increased haematological toxicity was not due to inclusion of elderly people in studies with more toxic agents. Renard and Aapro (unpublished data) looked at 146 ovarian cancer patients included in phase II studies of the EORTC (European Organization for Research and Treatment of Cancer) Early Clinical Trials Group. The patients received iproplatin, menogaril, or deoxyazacytidine. As shown in Table 10.1, there were no major differences for leukopenia but significantly more patients with thrombocytopenia among those aged above 60.

Pampallona (personal communication) looked at data from the Swiss Group for Clinical Cancer Research (SAKK) from patients treated with a combination of melphalan and cisplatin (Goldhirsch *et al*. 1988). There were only two patients above the age of 70 in this group of 158 patients. Patients above the age of 60 had a significantly higher median serum creatinine value than the younger patients, from the first cycle on. There

Table 10.1 Chemotherapy induced toxicity among ovarian cancer patients

	Age group		
	<60	60–9	≥70
ECOG patients		630	72
Haematological[a]		14%	26%
Infection		1%	3%
Vomiting		9%	6%
EORTC patients	94	42	10
WBC[a]	19%	13%	2%
Platelets[a]	30%	38%	60%
Infection	4%	1%	0%
Vomiting	31%	15%	3%
SAKK patients	108	48	2
WBC median		No significant difference	
Platelet median		No significant difference	
Creatinine		$p < 0.05$ for higher median among patients above 60	

[a] White blood cells (WBC) less than $2000/mm^3$, platelets less than $50\,000/mm^3$. For references see text.

was no statistically significant difference of median white blood cell or platelet nadirs.

Response to treatment has been reported by Begg and Carbone (1983) to be decreased (15 per cent) in elderly patients above 70 yers of age compared with younger patients (26 per cent responses, $p < 0.05$). Similarly they have shown that the duration of response is longer (53 weeks) in the younger patient group than in the older group (34 weeks). Again, it is unclear how comparable their two patient groups are and therefore we sought data from other European studies. In the phase II studies analysed by Renard and Aapro, the response rate was dismal and cannot be used for comparative purposes. Examining the reason for treatment discontinuation, there were more refusals to continue among patients aged 70 or more (40 per cent compared with 16 per cent in the younger group). The SAKK data reported by Pampallona show a 61 per cent response rate among 108 patients less than 60 years old treated with melphalan and cisplatin with a duration of 2.3 years, compared with a 56 per cent response rate and a duration of 1.9 years in the older patients. Dalesio (personal communication) looked at the results of a phase III EORTC Gynaecological Cancer Group study, which compared a cisplatin based combination with a carboplatin based combination. There is no difference between the observed and expected numbers of deaths in any age group (Table 10.2). The West Midlands

Table 10.2 Response and/or survival data from European studies of ovarian cancer patients

	Age group		
	<60	60–9	>70
EORTC patients	211	115	15
Dead/predicted	129/133	72/68	7/7
SAKK patients	108	48 + 2 above	70
Response	61%	56%	
Duration (median, in years)	2.3	1.9	
West Midlands	81	31 + 11 above	70
Alive	48%	39%	

For references see text.

groups study 3002 has been submitted to a preliminary analysis and there appeared to be a trend towards a worse survival among older patients (age limit 60, Table 10.3; Blackledge, personal communication).

Conclusion

There should be no 'chronological age' related bias against treatment of ovarian cancer in elderly people. It is to be hoped that gynaecological examinations in elderly people become more widespread, in order to detect patients with less advanced disease for surgery, which is best performed after careful patient preparation. The above data indicate that even among highly selected ovarian cancer patients there is a trend for increased toxicity from chemotherapy as age advances. This finding is not surprising, in view of the declining physiological reserves of aged people. However, as one can observe that the therapeutic results in the European study groups at least are not significantly different among various age groups, a recommendation for studies addressing specifically the use of curative and palliative chemotherapy in patients aged 70 years old or more seems justified.

Key points

- More than 50 per cent of patients with ovarian cancer are aged over 70.
- Age is not a determinant of outcome.

- Surgical mortality in selected patients can be as low as 3 per cent.
- There may be increased toxicity from chemotherapy with increase in age but little difference in efficacy.

References

Begg, C.B. and Carbone, P.P. (1983). *Cancer*, **52**, 1986–92.

Boring, C.C., Squires, T.S., and Tong, T. (1992). *Ca—Cancer Journal for Clinicians*, **42**, 19–33.

Castiglione, M., Gelber, R.D., and Goldhirsch, A. for the International Breast Cancer Study Group (1990). *Journal of Clinical Oncology*, **8**, 519–26.

Goldhirsch, A., Greiner, R., Dreher, E., Sessa, C., Krauer, F., Forni, M., Jjngi, F.W., Brunner, K.W., Veraguth, P., and Engeler, V. (1988). *Cancer*, **62**, 40–7.

Holmes, F.F. and Hearne, E. (1981). *Journal of the American Geriatrics Society*, **29**, 55–7.

Kennedy, A.W., Flagg, J.S., and Webster, K.D. (1989). *Gynecologic Oncology*, **32**, 49–54.

La Vecchia, C., Lucchini, F., Negri, E., Boyle, P., Maisonneuve, P., and Levi, F. (1992). *European Journal of Cancer*, **28**, 927–98.

Lawton, F.G. and Hacker, N.F. (1990). *Obstetrics and Gynecology*, **76**, 287–9.

Redman, J.R., Petroni, G.R., Saigo, P.E., Geller, N.L., and Hakes, T.B. (1986). *Journal of Clinical Oncology*, **4**, 515–23.

Samet, J., Hunt, W.C., Key, C., Humble, C.G., and Goodwin, J.S. (1986). *Journal of the American Medical Association*, **255**, 3385–90.

Yancik, R., Ries, L.G., and Yates, J.W. (1986). *American Journal of Obstetrics and Gynecology*, **154**, 639–47.

11

Acute myelogenous leukaemia

R. Zittoun, M. Baudard, and J.P. Marie

Age is a major prognostic factor in acute leukaemia (Cevreska and Gale 1987; Keating *et al.* 1988; Zittoun *et al.* 1989). This effect is observed at all ages, including childhood, and in all cytological types. In elderly patients, the prognosis of acute myelogenous leukaemia (AML), the most frequent cytologic type in this age group, is poor, both in terms of achievement of complete remission (CR) and overall survival (Buchner *et al.* 1985; Champlin *et al.* 1989; Preisler *et al.* 1987; Tucker *et al.* 1990; Wahlin *et al.* 1991; Yates *et al.* 1982). Age is an adverse prognostic factor for response, but once a CR is achieved, remission duration seems independent of age in most studies (Cevreska and Gale 1987; Keating *et al.* 1988). Some authors have published results from single centres with equivalent results for induction of CR to those in younger patients (Foon *et al.* 1981; Reiffers *et al.* 1980; Gale and Foon 1986) but most cooperative studies have shown that age *per se* correlates with a lower CR rate (Rees *et al.* 1986, Toronto Leukaemia Study Group 1986).

The concern about the poor prognosis of elderly AML is enhanced by the fact that there is an increased frequency of AML with age, with more than 50 per cent of cases being observed above 60 years of age (Brincker 1985; Cartwright and Staines 1992). This chapter addresses several questions: (1) what are the characteristics of elderly AML, (2) is there an increased risk of treatment-related morbidity and mortality, (3) are the leukaemic cells of elderly AML patients more resistant to cytotoxic drugs, (4) what are the results of specific treatment protocols for elderly AML, (5) what could be the therapeutic value of differentiating agents, and of haemopoietic growth factors, and (6) what are the psychosocial factors involved in the management and especially in medical decision-making?

Characteristics of elderly AML

There is an increased frequency of AML with ageing. Recent data from the USA show the incidence of acute leukaemia to be 5.7 cases per 100 000 population at age 4, 3.8 cases at age 50, and 23.0 cases at age 80 (National Cancer Institute 1987). Several other epidemiological studies have shown this increased incidence in the elderly (Brincker 1985; Tucker *et al.* 1990). The most frequent cytologic subtypes in elderly patients are of the AML group – myelocytic, monocytic (Whitely *et al.* 1990)—despite the fact that the increased frequency with age is also observed in adult acute lymphoblastic leukaemia (ALL), once the peak in childhood has been excluded. Some papers have suggested differences between AML in the elderly and in younger adults, such as lower incidence of Auer rods, lower haemoglobin levels, leucocyte counts, and bone marrow blasts (Beguin *et al.* 1985; Bloomfied and Theologides 1973).

AML can be divided into either *de novo* AML, or disease that occurs secondarily to myeloproliferative or myelodysplastic disorders, through clonal transformation ('blast crisis'). Elderly AML is characterized by a higher frequency of antecedent haematological disorders, which are adverse prognostic factors (Estey *et al.* 1982). In addition, in a proportion of cases aetiological factors can be identified in AML, such as prior occupation or a previous anticancer treatment, including especially alkylating agents. The leukaemogenic effect of these aetiological factors is dose dependent, with a cumulative activity. The time lag between the exposure to a leukaemogenic agent and the occurrence of AML is 2–10 years, with a peak incidence of AML at about 5 years (Kaldor *et al.* 1990). In addition, for those cases of AML secondary to occupational exposure, the time to the occurrence of AML can be more than 10 years (Yin *et al.* 1987). There is frequently a two-step development of secondary AML, with a first phase of pre-leukaemic abnormalities, especially a myelodysplastic syndrome such as a refractory anaemia with excess of blasts, followed by transformation into true AML (Crane and Keating 1991). All these observations may explain why AML is more frequent in an aged population.

There are two other major biological characteristics in elderly AML. Firstly, an increased frequency of specific cytogenetic abnormalities such as deletions of chromosomes 5 or 7, or trisomy 8 is observed (Li *et al.* 1983). These abnormalities are common features of myelodysplastic syndromes and secondary AML, and have been confirmed as adverse prognostic factors (Keating *et al.* 1988). Secondly, when CR are achieved in AML, they are generally non-clonal, with recovery of normal haemopoiesis, the cells from the three myeloid cell lines being polyclonal and different from the leukaemic clone. However, recent studies have shown that clonal remissions can be observed sometimes in AML, and such remissions are more frequent in elderly patients (Fialkow *et al.* 1987). One hypothesis is that there are less residual normal, non-clonal, haemopoietic stem cells left in elderly cases of AML, the malignant process having involved the very primitive stages of differentiation.

Another characteristic of AML in the elderly is the frequency of comorbidity, such as cardiovascular diseases, malignancy, and mental impairment. The total background of elderly AML is characterized by a marked vulnerability; the associated diseases may limit treatment possibilities because of an increased risk of toxicity, so that a relatively high proportion of patients are treated suboptimally. As an example, many patients have at presentation antecedent or persistent cardiac disorders, such as ischaemia, past myocardial infarction, or reduced left ventricular function, which limits the use of anthracycline drugs.

All these characteristics of elderly AML are adverse prognostic factors (Cevreska and Gale 1987; Keating *et al.* 1981; Reiffers *et al.* 1983) but age itself appears in several multivariate analyses as an indicator of poor response to induction therapy (Keating *et al.* 1988; Zittoun *et al.* 1989).

Increased risk of treatment-related toxicity and death

The major cause of failure to respond to induction treatment in elderly AML is the increased risk of death, mainly due to infection, following a conventional induction chemotherapy regimen (Keating *et al.* 1981; Zittoun *et al.* 1989). This increased risk is not limited to elderly patients. Age is rather a discrete variable, with a reduced mortality rate below 20, and then a progressively higher frequency in each decade of age. Yates *et al.* (1982) have published the results of a Cancer and Leukemia Group B (CALGB) study where the influence of age was assessed with three different induction regimens, combining cytosine arabinoside (Ara-C) with 3 days of either daunorubicin 45 mg/m^2, or daunorubicin 30 mg/m^2, or adriamycin 30 mg/m^2. Mortality from treatment was a major problem in elderly patients, occurring in more than 50 per cent in the age groups above 70, the least toxic regimen being daunorubicin 30 mg/m^2. As a consequence, the rate of CR decreased from more than 80 per cent to 20 per cent in patients respectively below 20 and above 70 with the higher dose of daunorubicin. At a dose of 30 mg/m^2, which was suboptimal in the younger patients, there was a 40 per cent CR rate in the older patients.

In our own experience at Hotel Dieu (Table 11.1), we have confirmed that there is a high risk of mortality in elderly patients. Of the 167 patients aged more than 60 treated between 1980 and 1989 who were given any kind of induction chemotherapy, the rate of toxic death increased from 19 per cent in patients of the age group 60–4 to more than 30 per cent in those aged more than 75. The higher rate of CR was observed in patients aged 60–4, then this rate declined to around 15 per cent in patients aged 70 or more. The proportion of cases surviving the induction period without entering into CR ('resistant patients') was higher than reported by the CALGB. This relatively lower rate of induction death and higher frequency of resistance in our experience can be explained by modified, and hence less toxic, induction treatments, the dose of daunorubicin being limited to 30 mg/m^2 on the basis of the CALGB experience, and by improved supportive therapy so that fewer patients died from infection or haemorrhage.

Although limited to patients aged up to 65 years old, our AML 6 European Organization for Research and Treatment of Cancer (EORTC) study has confirmed that age is a major prognostic factor for the results of induction treatment in AML (Zittoun *et al.* 1989). The two other major clinical prognostic factors were the WHO performance status and fever at presentation. In addition, an interdependence was observed between these prognostic factors, with increased risk of death in febrile patients of the older age groups.

In addition, besides the risk of death from infection during marrow hypoplasia, there is an increased risk of extra-haematological toxicity. As an example, age over 50 is the major risk factor for cerebellar toxicity following administration of high-dose Ara-C (Herzig *et al.* 1987).

Are the leukaemic cells of elderly patients more resistant to cytotoxic drugs?

Clinical studies have shown that one of the major causes of failure of induction treatment is death, either early before full administration of the cytotoxic drugs, or 'toxic death' during the hypoplastic phase. It is frequently difficult to evaluate the degree of sensitivity or resistance to the antileukaemic agents in patients dying during the induction period: residual disease can hardly be evaluated in dying patients, autopsy documents are scarce, and, as in our own series, increased resistance in the elderly could be due to suboptimal treatments aimed at reducing the risk of morbidity and mortality.

The question can be addressed by *in vitro* studies. Clonogenic assays are

Table 11.1 Elderly AML 1980–9 (Hotel Dieu): response to any induction treatment according to age (abstention excluded)

	60–4	65–9	70–4	75–9	≥80
n	47	25	49	33	13
CR	19 (40.4)	6 (24)	7 (14.2)	5 (15.1)	2 (15.3)
Death	9 (19.1)	6 (24)	7 (14.2)	11 (33.3)	4 (30.7)
Resistance	19 (40.4)	10 (40)	35 (71.4)	17 (51.5)	7 (53.8)

$p = 0.0003$ between patients of 70 or older versus patients younger than 70. The figures in parentheses are percentages.

currently performed to evaluate the *in vitro* proliferation of leukaemic stem cells (CFU-L), as well as their capacity of self-renewal following reseeding of the primary colonies. We have observed that primary colonies ('PE 1') can be obtained in most patients, and that the secondary colonies ('PE 2') are a major prognostic factor both for response to induction treatment and disease-free survival, patients with a higher PE 2 count being less likely to achieve a CR and more likely to have a shorter remission duration (Delmer *et al*. 1989). The suicide index, following *in vitro* exposure to tritiated thymidine (3H-TdR), is the easiest way to study the proportion of CFU-L in the S-phase of the cell cycle. Furthermore, the *in vitro* sensitivity of CFU-L to anthracycline drugs and Ara-C correlates very significantly with the clinical sensitivity or resistance to the induction treatment combining these two agents, thereby predicting the outcome of the induction therapy. We have therefore compared the results of these various *in vitro* parameters in AML patients aged up to or more than 60. Table 11.2 shows that there is no difference between these two age groups for all these parameters. There is a trend for a lower PE 2 in patients aged more than 60, but this difference – which could correspond to a paradoxically higher sensitivity of elderly patients—is not significant. Our results differ from those of Giannoulis *et al*. (1984) who showed an increased *in vitro* cell growth in elderly patients with AML, but using a different technique, examining blood instead of bone marrow.

In our study, the clonogenic cells of elderly patients were as sensitive *in vitro* to leukaemic agents as those from younger patients. Can we conclude from this single set of observations that leukaemic cells are no more resistant to cytotoxic drugs in elderly patients that in younger patients? Such an assumption should be made with caution, since some confounding factors could account for decreased sensitivity in elderly AML. This is especially true for the cytogenetic pattern, which is generally considered to be a major – if not the most important—prognostic factor in AML. As already emphasized, there is an increased frequency of adverse cytogenetic patterns in elderly AML, such as deletions of chromosome 5 or 7, trisomy 8, or complex abnormalities (Rowley 1981). All these abnormalities are highly correlated with lower response rate to induction treatment and shorter remission duration (Keating *et al*. 1988). Thus they could account in part for the lower sensitivity of elderly patients with AML to cytotoxic treatments. However, a different view has been taken by Nakamura *et al*. (1991) who, in a selected small series of AML patients with normal cytogenetics, observed an increased resistance to induction treatment in patients aged more than 65, which accounted for most induction failures.

Table 11.2 Increased resistance of leukemic cells of elderly AML patients to cytotoxic drugs? Clonogenic assays, Hotel Dieu

	PE 1		PE 2		Suicide		*In vitro* sensitivity (DNR + Ara-C)	
	n	m	n	m	n	m	n	m
⩾ 60 years	92	65+√−40	60	32+√−38	63	39+√−24	73	71+√−27
< 60 years	98	69+√−55	53	45+√−37	77	42+√−27	81	73+√−23
P		NS		0.08		NS		NS

NS not significant.

What can we learn from specific protocols for elderly patients?

A limited number of treatment protocols have been addressed specifically
to elderly cases of AML. The results of recent studies are presented in
Table 11.3, which shows great variability in treatments, ranging from very
conservative approaches with monochemotherapy, to intensive induction
treatments using conventional or even high-dose Ara-C. Most of these
studies were based on a limited number of patients, few were randomized,
and firm conclusions cannot be drawn. For most other studies performed
in AML, either the upper age limit of inclusion was 65 or 60, or there was no
age limit, but other exclusion criteria—such as a poor performance status
or comorbidity – could bias and optimize the true treatment results.

The first trial specifically for the elderly, performed by Kahn *et al*. (1984),
has confirmed the previous observations of CALGB, in that the optimum
therapeutic index in elderly patients was obtained with an attenuated dose
of chemotherapy, mainly from dose reduction of daunorubicin (DNR). The
higher dose resulted in increased toxicity and treatment-related mortality
with, as a consequence, a significantly longer overall survival with the
attenuated-dose induction treatment.

Some authors have adopted either low-dose Ara-C alone (Powell *et al*.
1989), or high-dose Ara-C (Lazarus *et al*. 1991). A popular induction
treatment in elderly AML patients is low-dose Ara-C (generally 10 mg/m^2
q. 12 hours for 21 days) (Cheson and Simon 1987). In a prospective
multicentre study comparing this regimen with a more standard treatment
combining rubidazone and conventional dose Ara-C, Tilly *et al*. (1990)
observed a lower CR rate with low-dose Ara-C (52 versus 32 per cent).
However, 22 per cent more patients achieved a partial response (PR)
with low-dose Ara-C, resulting in no significant difference between the
two regimens for the total response rate (CR + PR). In addition there
was no difference for the overall survival; some patients treated with
low-dose Ara-C were assumed to have a better quality of life, with less
toxicity and time spent in the hospital, although low-dose Ara-C gives rise
to appreciable haematological toxicity in most patients. Longer duration of
induction treatment with low-dose Ara-C, up to 42 days, did not result in a
higher CR rate, and caused a higher rate of induction deaths in the series
of Powell *et al*. (1989).

The value of high-dose Ara-C combined with DNR has been studied in
one small series of patients by Lazarus *et al*. (1991), with a relatively high
(43 per cent) CR rate, but also a high rate of induction death (53 per
cent). Similar observations have been made with intermediate-dose Ara-C
by Letendre *et al*. (1989).

Several options have been studied to replace the use of DNR for
remission induction. In the EORTC AML 9 trial, the combination of
mitozantrone with Ara-C resulted in a higher CR rate than the standard

Table 11.3 Specific clinical trials for elderly AML

Reference	Phase	Number of patients	Induction RX.	Results			CR duration (median)	Overall survival (median)
				CR	Death	Resistance		
Kahn et al. 1984	III	40	Full } DAT	25	60	15[a]		29 d[a]
			Attenuated	30	25	45[a]		150 d[a]
Powell et al. 1989	II	44	LD Ara-C × 42 d	23	25	53	9.9 m	3 m
Lazarus et al. 1989	I–II	21	HD Ara-C + DNR	43	38	19	9.0 m	6 m
Harousseau et al. 1989	II	20	Oral idarubicin	40	25	35		
Lowenberg et al. 1989	III	60	Immediate DNR–Ara-C }	58	10	32[a]		21 w[a]
			Delayed HU–Ara-C	0	62[b]	38[a]		11 w[a]
Letendre et al. 1989	II	30	ID Ara-C }	33	40	27	10.0 m	
			LD Ara-C × 21 d	32[c]	10	58	8.3 m	8.8 m
Tilly et al. 1990	III	87	RDZ + Ara-C }	52	31	17	13.8 m	12.8 m
			DNR + Arac-C	38	12	50		
Lowenberg 1991	III	488	MTZ + Ara-C }	52	16	32		7.5 m

LD, low dose; ID, intermediate dose; HD, high dose; RDZ, rubidazone; MTZ, mitozantrone; DAT, Daunorubicin, cytosine arabinoside, 6-thioguanine. d, day; w, week; m, month. [a]Significant difference. [b]8 of 18 deaths before commencement of chemotherapy. [c]+ 22 per cent >PR (versus 2 per cent in intensive care); no significant difference overall for CR duration survival.

DNR—Ara-C induction (Lowenberg *et al.* 1989). Harousseau *et al.* (1989) obtained 40 per cent CR in 20 patients treated with oral idarubicin alone.

Finally the question of whether elderly AML patients should be treated immediately, or should wait with supportive care and receive only palliative chemotherapy in case of progression, has been the basis of the AML 7 trial of EORTC. Only patients treated with immediate induction chemotherapy achieved a CR, and despite the short time lapse from progression to administration of chemotherapy in the 'wait and see' arm, there was a significantly shorter overall survival with this option which was expected to be more conservative (Lowenberg *et al.* 1989). However, as with most other studies in elderly AML, this was based on a limited number of patients, and could not avoid some criticisms about the methods used for palliative chemotherapy, together with the unusually high CR rate in the standard arm (Cassileth 1990).

The problem of the optimum post-remission treatment after achievement of CR is even more complex (Champlin *et al.* 1989). The CR duration is generally considered as independent of age (Cevreska and Gale 1987; Foon *et al.* 1981; Wahlin *et al.* 1991). However, in several studies shorter CR duration and/or disease-free survival (DFS) were observed in older patients (Rees *et al.* 1986; Zittoun *et al.* 1989). CR duration should be considered independently from DFS, since an increased risk of mortality during remission can be expected in elderly patients because of comorbidity factors and an increased risk of toxic death following consolidation—maintenance treatments. Even if CR duration appears to be shorter in older patients, this might be because age is an independent prognostic factor, or because some other parameters such as clonal remissions or unfavourable cytogenetics act as confounding factors.

There is a general tendency to consider that intensive post-CR treatments are the best way to prolong CR duration and to increase the rate of long-term remitters or cured patients. In young adults, such intensive cytoreductive treatments include high-dose chemotherapy, and allogeneic or autologous bone marrow transplantation (BMT). However, such treatments are hardly feasible in elderly patients: the limit of age put by most authors for allogeneic BMT is 45 years, and 45–60 years for autologous BMT. Older patients tolerate poorly the high-dose chemotherapy preparative regimens and have a higher incidence of graft versus host disease and other complications. High-dose chemotherapy such as Ara-C is difficult to repeat in elderly patients. Therefore in this age group standard consolidation—maintenance treatments seem to be the maximum tolerable, including one to three consolidation courses comparable with the induction treatment with some dose reduction, followed by maintenance cycles combining usually Ara-C and thioguanine. The value of such consolidation and/or maintenance treatments has been questioned repeatedly. However, recent randomized studies have shown that patients receiving no therapy beyond induction of CR have a median remission duration significantly shorter than patients receiving maintenance (Cassileth *et al.* 1988). Thus maintenance treatment seems beneficial in patients who cannot tolerate consolidation treatment.

The value of standard consolidation treatment preceding maintenance therapy is questionable; several randomized studies have shown that they provide a marginal benefit in patients receiving a subsequent maintenance treatment (Cassileth *et al.* 1984). Most elderly AML patients can tolerate only one to three courses of consolidation treatment (Champlin *et al.* 1989); each course will increase the risk of morbidity and toxic death, prolong the duration of hospitalization, and thus decrease the quality of survival. In the study from Tucker *et al.* (1990), where all AML patients received six consecutive courses of combined adriamycin, Ara-C and 6-thioguanine for induction and consolidation, treatment was often unfinished in patients over 60 on account of unacceptable toxicity, with only 2/88 patients receiving the planned six cycles of treatment. These various studies encouraged the EORTC Leukaemia Group to test the value of repeated courses of maintenance treatment with low-dose Ara-C.

The overall results of treatment in elderly AML are best described by the overall survival Kaplan—Meir estimates. In our experience, when we considered the survival duration of 235 consecutive AML patients over 60 treated in our department during the decade 1980–9, there was no difference between the patients entered during the first 5 years and those treated during 1985–9. By contrast, during the same period there was a trend for improvement of overall survival in younger patients. These results are similar to those from Whitely *et al.* (1990), who found no improvement of survival in the 1980s as compared with the 1970s.

Selection bias in results of trials on elderly AML

The results of most published studies could be considered as optimized, since they rarely avoid some selection bias. Such bias was clearly evidenced retrospectively when we analysed the rate of inclusion of AML patients in our department, where all the patients are supposed to enter into specific trials of the EORTC. Elderly patients were excluded from the AML trials of EORTC until 1983, the inclusion criteria allowing only patients up to 65 years old. After that date, the EORTC Leukaemia Group started designing specific protocols for elderly AML, the age limit being lowered to 60 years. Three consecutive protocols were activated. From December 1983 to March 1986, the AML 7 protocol assessed the value of a 'wait and see' policy followed by palliative chemotherapy. From April 1986 to September 1990 the AML 9 protocol compared mitozantrone—Ara-C with DNR—Ara-C for induction, and low-dose Ara-C for maintenance versus

Acute myelogenous leukaemia

no maintenance (see above). Since October 1990, the AML 11 protocol has assessed the value of granulocyte macrophage colony stimulated factor (GM-CSF) during induction, starting 24 hours before a DNR – Ara-C course, and continuing until day 28 or neutrophil recovery.

Of 235 AML patients aged more than 60 years, referred to our institution during the decade 1980–9, 79 per cent were treated with chemotherapy, whether palliative or more intensive, and 21 per cent did not receive any form of chemotherapy, either at presentation or subsequently. The main reasons for witholding chemotherapy were poor general condition (16), cardiac (6), infection (6), respiratory (5), refusal (4), haemorrhage (2), psychic (2), various (3), and unspecified in 24 patients. Looking at the 1983–9 period corresponding to the activation of protocols for elderly AML, 32 per cent only of the 169 patients consecutively hospitalized were included in the EORTC trials (during the same period the rate of inclusion of patients below 60 years was 83 per cent!). The main reasons for non-inclusion of elderly AML patients were poor general condition (13), cardiac (12), respiratory (6), renal (6), infection/haemorrhage (3), refusal (2), various (8), and unspecified (65!).

When we compared the main data and the clinical outcome of patients included or not included in these specific protocols for elderly AML, the two major differences were age and performance status, patients included being significantly younger and having a better performance status (Table 11.4). Other parameters such as haemoglobin, blood counts (White blood cells (WBC) and platelets) and Lactic dehydrogenase (LDH) were similar

Table 11.4 Inclusion of elderly AML in EORTC protocols, Hotel Dieu 1983–9

	Inclusion	Non-inclusion	p
Age	$69.9 + \surd -6.3$	$73.4 + \surd -7.6$	0.006
WHO > 2	10/50 (20)	49/81 (60)	0.001
WBC	$52.7 + \surd -66$	$60 + \surd -100$	NS
Haemoglobin	$8.8 + \surd -2$	$9.2 + \surd -1.9$	NS
Platelets	$99 + \surd -87$	$96 + \surd -133$	NS
LDH	$703 + \surd -839$	$1160 + \surd -1533$	NS
Response to treatment			
CR	20 (37)	8 (7.5)	<0.02
Death	10 (18.5)	36 (34)	<0.001
Resistance	24 (44.5)	50 (47)	NS
Lost to follow-up	0	12 (11.5)	–
Mean duration follow-up (days)	274	139	–

The figures in parentheses are percentages. NS, not significant.

in the two groups. The outcome could be considered only for those patients who were not included and who received any form of chemotherapy. The CR rate was far lower, and the death rate during induction higher. In addition the mean duration of follow-up was shorter (139 instead of 274 days) and 11.5 per cent of such patients were lost to follow-up (as against none of the patients included in protocols). The overall survival was also significantly shorter ($p < 0.0001$), with a median survival of 2 months compared with 6 months.

Our data are in agreement with recent papers in the literature, showing the impact of selection bias in elderly AML (Brincker 1985; Toronto Leukaemia Group 1986; Wahlin *et al.* 1991). The Danish and Canadian studies, especially, led to the conclusion that most published series are highly selected, and that the overall CR rates reported are much higher than they would have been if a substantial number of patients had not been excluded from treatment because of old age, poor general condition, and restrictive inclusion criteria into protocols (Brincker 1985; Toronto Leukaemia Study Group 1986). In addition it is assumed that many elderly AML patients are not referred to haematological units and are thus undertreated (Wahlin *et al.* 1991). The data of Wahlin *et al.*, showing better CR rates and survival in elderly AML treated with aggressive induction—consolidation than in those treated with low-dose regimens, favour the idea that performance status is a more important factor than age *per se* for the ability to survive such aggressive treatments (Wahlin *et al.* 1991). It is difficult to conclude from such studies whether the difference in response rate and survival according to the intensity of treatment is due to the treatment *per se*, or to selection criteria resulting in different allocations of treatment according to age, performance status, and/or comorbidity. One can assume, at the present time, that more intensive treatment should be reserved for elderly patients with a good performance status, although the dose of intercalating agents should be attenuated, and high-dose Ara-C should be avoided, as underlined above.

What is the role of differentiating agents and growth factors in elderly AML?

Differentiating agents

The capacity of various agents of different chemical families such as dimethyl sulphoxide (DMSO), vitamin A or D derivatives, low-dose cytotoxic agents, and biomodulators, to induce differentiation of acute leukaemia cells has been demonstrated in various experimental models in animals, as well as in human leukaemic cell lines (Koeffer 1983). The possible advantages of such a therapeutic approach are obvious, since it would replace cytotoxic treatments with their unavoidable killing effect on residual normal haematopoietic cells by reversal of the leukaemic process.

However, the first attempts, mainly based on low-dose cytotoxic agents, and especially low-dose Ara-C, were rather disappointing. Not only was the CR rate lower than with conventional dose cytotoxic drugs, but also these CR were rarely achieved through an apparent progressive differentiation of leukaemic cells without an aplastic phase (Cheson and Simon 1987).

More recently, the therapeutic use of a differentiating agent appeared promising in a specific subtype of AML, i.e. acute promyelocytic leukaemia. In this subtype, oral administration of all-trans-retinoic acid induces CR through a differentiating effect on the promyelocytic cells in a relatively high percentage of cases (Castaigne *et al.* 1990; Warell *et al.* 1991). Such CR are obtained through maturation of the leukaemic clones without a transient phase of bone marrow hypoplasia. As a consequence, the rate of toxic death is markedly reduced and is not related to infection or haemorrhage. However, this therapeutic effect is only obtained in this small subgroup of AML, characterized by a chromosomal translocation t15; 17 with expression of an aberrant retinoic acid-alpha nuclear receptor, which accounts for only a small minority of AML, especially in the elderly.

Growth factors

Haemopoietic growth factors, particularly granulocyte colony stimulating factor (G-CSF) and GM-CSF are now used widely to accelerate the recovery of neutrophils following intensive cytotoxic treatments for solid tumours or bone marrow transplantation. A similar therapeutic effect would appear to be even more useful in AML, where most toxic deaths are due to neutropenic infection during the post-induction or consolidation phases. Clinical trials started with some caution since leukaemic myeloblasts have receptors for GM-CSF and G-CSF, and are frequently stimulated *in vitro* to proliferate with these growth factors. Recent pilot studies have confirmed that administration after chemotherapy of GM-CSF in AML accelerates the recovery of polymorphonuclears, thus reducing the rate of infections and toxic deaths when compared with historical controls (Bettelheim *et al.* 1991; Buchner *et al.* 1991). These studies encouraged the activation of randomized prospective trials to assess the possible advantages, disadvantages, and the side-effects of G-CSF and GM-CSF; elderly AML patients have been selected as an appropriate group for such trials, taking into account the increased risk of toxic death during such intensive treatments. Special attention is paid to the potential stimulation of the leukaemic clones by GM-CSF. This could be an advantage, by recruiting more leukaemic cells into the cell cycle and increasing their sensitivity to the cytotoxic drugs, when the growth factor is administered before and during the combination of cytotoxics (Bettelheim *et al.* 1991). However, the risk of stimulating leukaemic regrowth when GM-CSF is administered or continued after completion of the cytotoxic courses was not observed in these two first pilot studies: only one patient out of 36 in the study from Buchner *et al.* (1991) had a marked leukaemic regrowth which was completely reversible after stopping GM-CSF.

Supportive and palliative care

Important progress has been made in the supportive care of AML, mainly from preventive and curative treatments of infections, by reducing the microbial content of the environment and of diet, and, in the case of overt or suspected infections, by the use of potent antibiotics including third-generation cephalosporins, vancomycin, and quinolones. Haemorrhages, which are the second cause of death, are efficiently prevented by platelet transfusions. This improvement in supportive care could explain why patients entered in more recent phases of clinical trials have a better survival than those entered some years earlier (Rees *et al.* 1986). This improvement has several consequences which should be stressed.

1. The survival of patients with slowly proliferative AML can be prolonged mainly through good supportive care. In such patients, cytotoxic treatments can be avoided for prolonged periods of time, or limited to low dose (or short courses) in order to slow down the leukaemic proliferation. However, the exact frequency of the smouldering acute leukaemias, although not clearly determined, is probably low, which is why most patients will die from progression of their disease despite good supportive care. The characteristics of these smouldering leukaemias are not yet clearly established, although some studies have pointed out that they are more frequently oligoblastic, of FAB M2 type, and with a lower labelling index (Van Slyck *et al.* 1983).
2. The economic cost of supportive care is high. Even when treated as out-patients, with supportive care being provided at home or in day care clinics, this is disproportionately more expensive than the cost of cytotoxic drugs alone.

The role and application of palliative care in patients with acute leukaemia is currently underestimated. Not only is classical bone pain frequent in AML, but also sore mouth and painful rectal abcesses. Acute infections can cause distressing symptoms, either acute dyspnoea in the case of lung infection, or abdominal pain and diarrhoea as a result of intestinal gram-negative infection. Malnutrition and fatigue are especially frequent. Death can occur in various conditions, some of them being very distressful.

A clear definition of prognosis is an absolute prerequisite for providing adequate palliative care. As with other diseases, especially malignancies, this should be aimed at relief and prevention of pain and suffering. One should be aware, however, of the specific problems of AML such as an increased risk of acute and painful complications as a terminal event, and for elderly patients the risk of physical and psychological degradation.

Psychosocial aspects of elderly AML

As shown by Yancik (1986), there are striking parallels between the conditions of old age and cancer: body integrity is reduced by physical disability and impairment, with frequent comorbidity. Aged persons and many cancer patients experience similar constraints in their physical mobility and daily routines. There is an increased financial and social dependence, with loss of autonomy and role. In addition, according to Yancik, persons of both groups are characterized by depersonalization, low self-esteem, loneliness, and isolation, and may not be able to remedy such circumstances. Consequently they often require a transfer of some power over their own life to others. Naturally, these changes are even more pronounced in elderly cancer patients. Furthermore, the relatively high prevalence of cognitive dysfunction in the elderly can increase this loss of role and self-power (Holland and Massie 1987) or encourage the relatives and caregivers to impose their own decisions and power.

These general features of malignancies in elderly patients generate even more dramatic situations in such acute, life-threatening diseases as AML. In most cases the diagnosis will lead to immediate hospitalization with social disruption, and administration of cytotoxic treatments with a serious risk of physical and psychosocial consequences. The statement that elderly AML patients should be treated as younger patients is questionable, since all studies have largely demonstrated an increased risk of treatment-related morbidity and mortality.

Medical decision-making should be based primarily on clinical judgement and patients' preferences, whereas other considerations such as subsumed quality of life values, and external factors including costs of medical care, or research and teaching needs or medicine, are accessory (Siegler 1982). Unfortunately, in an acute condition such as AML, determination of the patient's preferences is generally difficult, if one wants to avoid increasing his/her anxiety by raising the level of uncertainty about the prognosis of the disease and the outcomes of treatment. In addition, one should take into account that older patients with life-threatening diseases may prefer a therapeutic option with a lower long-term survival but a decreased risk of immediate death from the treatment (McNeil *et al*. 1978).

Conclusion

AML in the elderly is mainly characterized by an increased risk of treatment-related morbidity and mortality. Age *per se* is an independent, although not fully understood, adverse prognostic factor; however, performance status and comorbidity are more important than age when one considers the principle and intensity of a cytotoxic antileukaemic treatment, which remains up to now the best chance of achieving a CR and prolonging survival. The dilemma about the principle and intensity of cytotoxic treatment in elderly AML (Coppelstone *et al* 1989; Crosby 1968) cannot yet be concluded easily. The patient's preferences and values should also be taken into account, but can be difficult to elicit, and a more global approach of his/her psychosocial background is necessary in order to select the optimal therapeutic option.

The same holds true in the later course of the disease, after failure of treatments or at the time of relapse, when difficult decisions must be weighted such as continuation or discontinuation of supportive care—such as antibiotics and transfusion—and selection of the optimum environment for appropriate palliative and terminal care.

There is an increased agreement in the medical community to develop specific prospective trials and advanced age should not be an exclusion criteria. The unacceptable toxicity of very intensive treatment in elderly AML cases should lead to the development of protocols stratified according to age-groups, or to specific trials for AML in the elderly.

Key points

- Acute myelogenous leukaemia is increasing in frequency among the elderly.
- AML more frequently follows other haematological disorders and thus may have a worse prognosis in the elderly.
- Treatment toxicity increases progressively in patients aged over 80 years.
- The optimal therapeutic index in the elderly occurs after drug attenuation.
- It is important that the preferences or values of the patient are considered when making decisions on therapy.

- Prospective trials are required which do not exclude patients on age alone.

References

Beguin, Y., Bury, J., Fillet, B., and Lennes, G. (1985). Treatment of acute nonlymphocytic leukaemia in young and elderly patients. *Cancer*, **56**, 2587–92.

Bettelheim, P., Valent, P., Andreef, M., Tafuri, A., Haimi, J., Gorischek, C., Muhm, M., Sillaber, C., Haas, O., and Vieder, L. (1991). Recombinant human granulocyte-macrophage colony-stimulating factor in combination with standard induction chemotherapy in de novo acute myeloid leukaemia. *Blood*, **77**, 700–11.

Bloomfield, C.D. and Theologides, A. (1973). Acute granulocytic leukaemia in elderly patients. *J. Am. Med. Assoc.*, **226**, 1190–3.

Brincker, H. (1985). Estimate of overall treatment results in acute nonlymphocytic leukaemia based on age-specific rates of incidence and of complete remission. *Cancer Treat. Rep.*, **69**, 5–11.

Buchner, Th., Urbanitz, D., Hiddemann, W., Ruhl, Ludwig, W.D., Fischer, J., Aul, H.C., Vaupel, H.A., Kuse, R., and Zeile, G. (1985). Intensified induction and consolidation with or without maintenance chemotherapy for acute myeloid leukaemia (AML) : two multicenter studies of the German AML Cooperative Group. *J. Clin. Oncol.*, **3**, 1583–9.

Buchner, Th., Hiddemann, W., Koenigsmann, M., Zuhlsdorf, M., Wormann, B., Boeckmann, A., Freire, E.A., Innig, G., Maschmeyer, G. and Ludwig, W.D. (1991). Recombinant human granulocyte-macrophage colony-stimulating factor after chemotherapy in patients with acute myeloid leukaemia at higher age or after relapse. *Blood*, **78**, 1190–7.

Cartwright, R.A. and Staines, A. (1992). Acute leukaemias. In *Epidemiology of haematological disease*, Part I, Vol. 5, (ed. A.F. Fleming), pp. 1–26. (ed). Bailliere's Clinical Haematology. London.

Cassileth, P.A. (1990). Intensive therapy in elderly patients with acute myeloid leukaemia. *J. Clin. Oncol.*, **8**, 937.

Cassileth, P.A., Begg, C.B., Bennett, J.M., Bozdech, M., Kahn, Sb., Weiler, C., and Glick, J.H. (1984). A randomised study of the efficacy of consolidation therapy in adult acute nonlymphocytic leukaemia. *Blood*, **63**, 843–7.

Cassileth, P.A., Harrington, D.P., Hines, J.D., Oken, M.M., Mazza, J.J., McGlave, P., Bennett, J.M., and O'Connell, M.J. (1988). Maintenance chemotherapy prolongs remission duration in adult acute nonlymphocytic leukaemia. *J. Clin. Oncol.*, **6**, 583–7.

Castaigne, S., Chomienne, Ch., Daniel, M.T., Ballerini, P., Berger, R., Fenaux, P., *et al.* (1990). All-trans retinoic acid as a differentiation therapy for acute promyelocyte leukaemia: I, clinical results. *Blood*, **76**, 1704–9.

Cevreska, L., and Gale, R.P. (1987). Prognostic factors in acute myelogenous leukaemia. *Hematol. Blood Transfus.*, **30**, 376–9.

Champlin, R.E., Gajewski, Jl., and Golde, D.W. (1989). Treatment of acute myelogenous leukaemia in elderly. *Semin. Oncol.*, **16**, 51–6.

Cheson, B.D. and Simon, R. (1987). Low-dose ara-C in acute nonlymphocytic leukaemia and myelodysplastic syndromes: a review of 20 years experience. *Semin. Oncol.*, **14**, 126–33.

Copplestone, J.A., Smith, A.G., Osmond, C., Oscier, D.G., and Hamblin, T.J.

(1989). Treatment of acute myeloid leukaemia in the elderly: a clinical dilemma. *Hematol. Oncol.*, **7**, 53–9.

Crane, M.M. and Keating, M.J. (1991). Exposure histories in acute nonlymphocytic leukaemia patients with a prior preleukaemia condition. *Cancer*, **67**, 2211–14.

Crosby, W.H. (1968). To treat or not to treat acute granulocytic leukaemia. *Arch. Intern. Med.*, **122**, 79–80.

Delmer, A., Marie, J.P., Thevenin, D., Cadiou, M., Viguie, F., and Zittoun, R. (1989). Multivariate analysis of prognostic factors in acute myeloid leukaemia: value of clonogenic leukaemia cell properties. *J. Clin. Oncol.*, **7**, 738–46.

Estey, E.H., Keating, M.J., McCredie, K.B., Bodey, G.P., and Freireich, E.J. (1982). Causes of initial remission induction failure in acute myelogenous leukaemia. *Blood*, **60**, 309–15.

Fialkow, P.J., Singer, J.W., Raskind, W.H., Adamson, J.W., Jacobson, R.J., Bernstein, I.D., Dow, L.W., Najfeld, V., and Veith, R. (1987). Clonal development, stem-cell differentiation, and clinical remissions in acute nonlymphocytic leukaemia. *N. Engl. J. Med.*, **317**, 468–73.

Foon, K.A., Zighelboim, J., Yale, C., and Gale, R.P. (1981). Intensive chemotherapy is the treatment of choice for elderly patients with acute myelogenous leukaemia. *Blood*, **58**, 467–70.

Gale, R.P. and Foon, K.A. (1986). Acute myeloid leukaemia: recent advances in therapy. *Clin. Haematol.*, **15**, 781–810.

Giannoulis, N., Ogier, C., Hast, R., Lindblom, B., Sjogren, A.M., and Reizenstein, P. (1984). Difference between young and old patients in characteristics of leukaemia cells. *Am. J. Hematol.*, **16**, 113–21.

Harousseau, J.L., Rigal-Huguet, F., Hurteloup, P., Guy, H., Milpied, N., and Pris, J. (1989). Treatment of acute myeloid leukaemia in elderly patients with oral idarubicin as a single agent. *Eur. J. Haematol.*, **42**, 182–5.

Herzig, R.H., Hines, J.D., Herzig, G.P., Wolff, S.N., Cassileth, P.A., and Lazarus, H.M. (1987). Cerebellar toxicity with high-dose cytosine arabinoside. *J. Clin. Oncol.*, **5**, 927–32.

Holland, J.C. and Massie, M.J. (1987). Psychosocial aspects of cancer in the elderly. *Clin. Geriatr. Med.*, **3**, 533–9.

Kahn, S.B., Begg, C.B., Mazza, J.J., Bennett, J.M., Bonner, H., and Glick, J.H. (1984). Full dose versus attenuated dose daunorubicin, cytosine arabinoside and 6-thioguamine in the treatment of acute nonlymphocytic leukaemia in the elderly. *J. Clin. Oncol.*, **2**, 865–70.

Kaldor, J.M., Day, N.E., Clarke, E.A., van Leeuwen, F.E., Henry-Amar, M., Fiorentino, M.V., Bell, J., Pedersen, D., Band, P., and Assouline, D. (1990). Leukaemia following Hodgkin's disease. *N. Engl. J. Med.*, **322**, 7–13.

Keating, M.G., McCredie, K.B., Benjamin, R.S., Bodey, G.P., Zander, A., Smith, T.L., et al. (1981). Treatment of patients over 50 years of age with acute myelogenous leukaemia with a combination of rubidazone and cytosine arabinoside, vincristine and prednisone (ROAP). *Blood*, **58**, 584–91.

Keating, M.J., Smith, T.L., Kantarjian, H., Cork, A., Walters, R., Trujillo, J.M., MeCredie, K.B., Gehan, E.A., and Freireich, E.J. (1988). Cytogenetic pattern in acute myelogenous leukaemia: a major reproductible determinant of outcome. *Leukemia*, **2**, 403–12.

Koeffer, H.P. (1983). Induction of differentiation of human acute myelogenous leukaemia cells : therapeutic implications. *Blood*, **62**, 709–21.

Lazarus, H.M., Vogler, Wr., Burnus, P., and Winton, E.F. (1991). High-dose cytosine arabinoside and daunorubicin as primary therapy in elderly patients with acute myelogenous leukaemia. A phase I–II study of the Southeastern Cancer Study Group. *Cancer*, **63**, 1055–9.

Letendre, L., Niedringhaus, R.D., Therneau, T.M., Gastineau, D.A., Goldberg, J.B., Maillard, J.A., Gundlach, W.T., Kardinal, C.G., and Pierre, R.V. (1989). Treatment of acute nonlymphocytic leukaemia in the elderly with intermediate high-dose cytosine arabinoside. *Med. Pediatr. Oncol.*, **17**, 79–82.

Li, Y.S., Khalid, G., and Hayhoe, F.G.J. (1983). Correlation between chromosomal pattern, cytological subtypes, response to therapy and survival in acute myeloid leukaemia. *Scand. J. Haematol.*, **30**, 265–77.

Lowenberg, B., Zittoun, R., Kerkhofs, H., Jehn, U., Abels, J., Debusscher, L., Cauchie, C., Peetermans, M., Solbu, G., and Suciu, S. (1989). On the value of intensive remission induction chemotherapy in elderly patients of 65+ years with acute myeloid leukaemia: a randomised phase III study of the European Organisation for Research and Treatment of Cancer Leukaemia Group. *J. Clin. Oncol.*, **7**, 1268–74.

McNeil, B., Weichselbaum, R., and Pauker, S.G. (1978). Fallacy of the five-year survival in lung cancer. *N. Engl. J. Med.*, **299**, 1397–1401.

Nakamura, H., Sadamori, N., Sasagawa, I., Itoyama, S., Tokunaga, S., Mine, M., and Ichimaru, M. (1991). Acute nonlymphocytic leukaemia with normal karyotype. Is its in vivo drug susceptiblity age-dependent? *Cancer Genet. Cytogene*, **51**, 67–71.

National Cancer Institute (1987). *Annual cancer statistics review*, National Institutes of Health, NIH Publication. No 88,2789, p. 192. National Cancer Institute, Bethesda, MD.

Powell, B.L., Capizzi, R.L., Muss, H.B., Bearden, J.D., Lyerly, E.S., Rosenbaum, D.L., Morgan, T.M., Richards, F., Jackson, D.V., and White, D.R. (1989). Low-dose Ara-C therapy for acute myelogenous leukaemia in elderly patients. *Leukemia*, **3**, 23–8.

Preisler, H., Davis, R.B., Kirshner, J., Dupre, E., Richards, III, F., Hoagland, H.C., Kopel, S., Levy, R.V., Carey, R., and Schulman, P. (1987). Comparison of three remission induction regimens and two postinduction strategies for the treatment of acute nonlymphocytic leukaemia: a Cancer and Leukaemia Group B Study. *Blood*, **69**, 1441–9.

Rees, J.K.H., Gray, R.G., Swirski, D., and Hayhoe, F.G.J. (1986). Principal results of the Medical Research Council's 8th acute myeloid leukaemia trial. *Lancet*, **ii**, 1236–41.

Reiffers, J., Raynal, F., and Broustet, A. (1980). Acute myeloblastic leukaemia in elderly patients. *Cancer*, **45**, 2816–20.

Reiffers, J., Bernard, Ph., David, B., Vezon, G., and Broustet, A. (1983). Acute myeloid leukaemia in elderly patients: results of chemotherapy and prognostic factors. *Haematologica*, **68**, 214–25.

Rowley, Jd. (1981). Association of specific chromosomal abnormalities with type of acute leukaemia and with patient age. *Cancer Res.*, **41**, 3407–10.

Siegler, M. (1982). Decision-making strategy for clinical–ethical problems in medicine. *Arch. Intern. Med.*, **142**, 2178–9.

Tilly, H., Castaigne, S., Bordessoule, D., Casassus, P., Le Prise, P.Y., Tertian, G., Desablens, B., Henry-Amar, M. and Degos, L. (1990). Low-dose Cytarabine versus intensive chemotherapy in the treatment of acute nonlymphocytic leukaemia in the elderly. *J. Clin. Oncol.*, **8**, 272–9.

Toronto Leukaemia Study Group. (1986). Results of chemotherapy for unselected leukaemia: effect of exclusion on interpretation of results. *Lancet*, i, 786–8.

Tucker, J., Thomas, A.E., Gregory, W.M., Ganesan, T.S., Malik, S.T.A., Amess, J.A.L., Lim, J., Willis, L., Rohatiner, A.Z., and Lister, T.A. (1990). Acute myeloid leukaemia in elderly adults. *Hematol. Oncol.*, **8**, 13–21.

Van Slyck, E.J., Reuck, J.N., Waddell, C.C., Janakiraman, N. (1983). Smouldering acute granulocytic leukaemia. Observations on its natural history and nonlymphocytic characteristics. *Arch. Intern. Med.*, **143**, 37–40.

Wahlin, A., Hornsten, P., and Jonsson, H. (1991). Remission rate and survival in acute myeloid leukaemia : impact of selection and chemotherapy. *Eur. J. Haematol.*, **46**, 240–7.

Warrell, R.P., Jr., Frankel, S.R., Miller, Wh., Jr., Scheinberg, D.A., Itri, L.M., Hittelman, W.N., Vyas, R., Andreeff, M., Tafuri, A., and Jakubowski, A. (1991). Differentiation therapy of acute promyelocytic leukaemia with tretinoin (all-trans-retinoic acid). *N. Engl. J. Med.*, **324**, 1385–3.

Whitely, R., Hannah, P., and Holmes, F. (1990). Survival in acute leukaemia in elderly patients. No improvement in the 1980s. *J. Am. Geriatr. Soc.*, **38**, 527–30.

Yancik, R. (1986). Parallels between research approaches to address problems of quality of life issues in the elderly and cancer patients. In *Assessment of quality of life and cancer treatment*, (ed. V. Ventrafridda, F.S.A.M. van Dam, R. Yancik, and M. Tamburini), pp. 51–64. Excerpta Medica, Amsterdam.

Yates, J., Glidewell, O., Wiernik, P., Cooper, M.R., Steinberg, D., Dosik, H., *et al.* (1982). Cytosine arabinoside with daunorubicin or adriamycin for therapy of acute myelocytic leukaemia : a CALGB study. *Blood*, **60**, 454–62.

Yin, S.N., Li, G.L., Tain, F.D., Fu, Z.I., Jin, C., Chen, Y.J., Luo, S.J., Ye, P.Z., Zhang, J.Z., and Wang, G.C. (1987). Leukaemia in benzene workers : a retrospective cohort study. *Br. J. Ind. Med.*, **44**, 124–8.

Zittoun, R., Jehn, U., Fiere, D., Haanen, C., Loewenberg, B., Willemze, R., Abels, J., Bury, J., Peetermans, M., and Hayat, M. (1989). Alternating v repeated postremission treatment in adult acute myelogenous leukaemia: a randomised phase III study (AML 6) of the EORTC Leukaemia Cooperative Group. *Blood*, **73**, 896–906.

12

Non-Hodgkin's lymphoma

Umberto Tirelli and Silvio Monfardini

Introduction

Older adults are at higher risk of developing and dying from malignant tumours than their younger counterparts, according to the latest Cancer Statistics Review 1973–87, reporting data from the USA.[1] Americans 65 years of age or older are ten times more likely to develop cancer than those under 65 years of age. In general, survival also decreases as age increases. In contrast to the decline of 4.5 per cent in the cancer mortality for persons under age 65 in the period 1973–87, the overall cancer mortality for persons aged 65 or older has increased by 13 per cent during this 15 year period. In patients aged over 65 those cancers with an increasing mortality include lung, brain, melanoma, multiple myeloma, kidney, oesophagus, ovarian, prostate, breast, pancreas, larynx, and non-Hodgkin's lymphoma (NHL). In contrast there has been a reduction in mortality for those with tumours of the liver, colorectum, uterus, bladder, thyroid, stomach, cervix, Hodgkin's disease, testis, and oral cancer. There has been a 55 per cent increase in NHL mortality during the period 1973–87 in patients older than 65 years of age.[1]

In Western countries, more than half of all cancer cases are diagnosed in persons aged 65 years or more, and approximately one-third in persons aged 70 years or older. These figures are likely to increase in the following decades as the population grows older.[2] Patients older than 70 years are generally excluded from therapeutic protocols because of their age and, furthermore, very few protocols specifically designed for elderly patients are available. As a consequence, there is little scientific knowledge on natural history, suitable diagnostic procedures, and efficacy of chemotherapy, radiation therapy, and surgery on cancer in old age.[3,4] Moreover, surprisingly little is known about the efficacy, toxicity, pharmacokinetics, and pharmacodynamics of chemotherapeutic agents in elderly cancer patients.

Approximately 25–35 per cent of cases of NHL occur in patients 70 years or older. Intermediate and high-grade histology NHL are observed with similar frequency in patients younger and older than 70 years of age,[4,5] whereas in elderly patients extranodal sites of involvement are more common than in younger patients.

There are several potential causes of age bias in both conducting and reporting on trials in NHL. The following are the most common. Firstly, the median age of the patient populations of series reported in the literature is usually between 50 and 55 years, sometimes between 45 and 50 years or 55 and 60 years, but rarely with a median age of more than 60 years. However, one-third of NHL patients are more than 70 years old and two-thirds are over 65. Secondly, recently reported clinical trials where an age of over 70 was not an exclusion criterion included the statement that there was no upper age limit for entry into the studies. In practice, the number of patients over 70 is small and the median age usually ranges between 50 and 55 years. Therefore, this statement does not mean that the conclusions reached in the studies are applicable to elderly NHL patients. Thirdly, patients may be grouped, for example, into those younger or older than 60 years, and complete response and survival rates compared. However, the median age of the older patients is usually not reported and hence the conclusion that older age does not influence complete response or survival rate is not acceptable. Fourthly, 'age is not a prognostic factor' is another common assertion; however, patients tend to be selected for entry into the study mainly because of their age. Fifthly, conclusions almost never state that the results presented are valid for a patient population of that median age. As a result, unquestioning use of the conclusions of trials in older patients may lead to an increased percentage of treatment-related toxic deaths.[6] The aim of this chapter is to report the results of prospective trials of treatment in elderly patients with NHL, and to discuss the problems of toxicity and the new directions for clinical research.

Selected prospective clinical trials

The results obtained in the prospective treatment of NHL of the elderly by four groups (Vancouver, Canada; Nebraska, USA; GELA (French group for the study of lymphomas) France; Aviano, Italy) are summarized.

The Vancouver group used two brief, weekly chemotherapy regimens designed specifically for elderly patients. Eligible patients were aged 65–85 years and had advanced stage diffuse large-cell lymphoma. A study of low-dose ACOP-B (doxorubicin, cyclophosphamide, vincristine, bleomycin, and prednisone) accrued 40 patients between March 1983 and September 1985; 65 per cent achieved complete response (CR), there were two toxic deaths, the actuarial failure-free survival was 19 per cent, disease specific survival 30 per cent, and overall survival 28 per cent with a maximum follow-up of 6 years. The VABE regimen (etoposide, doxorubicin, vincristine, bleomycin, and prednisone) accrued 32 patients between July 1985 and June 1987; 63 per cent achieved CR, there were two toxic deaths, the actuarial failure-free survival was 34 per cent disease specific survival 45 per cent, and overall survival 36 per cent with a maximum follow-up of 4 years.

The two study regimens had a different toxicity profile, low-dose ACOP-B being better tolerated and VABE having more considerable haematological toxicity. These two regimens have the advantage of brevity, out-patient administration, and acceptable long-term survival despite moderate toxicity. The authors state that low-dose ACOP-B should be selected to treat an elderly patient with an advanced stage large-cell lymphoma since this regimen was much better tolerated that VABE and since the long-term results showed a 30 per cent disease specific survival at 6 years.[7]

The Nebraska group treated 157 patients with diffuse aggressive NHL between September 1982 and May 1986 with cyclophosphamide, adriamycin, procarbazine, bleomycin, vincristine, and prednisone (CAP/BOP). Patients in this study ranged in age from 15 to 91 years with 112 patients 60 or more years old ($p = 0.01$). Deaths attributable to tumour or treatment-related toxicity were similar in both age groups. The difference in survival was due to other causes of death not obviously related to the lymphoma or its therapy, occurring in 22 per cent of patients 60 years of age or more but in only 2 per cent of patients less than 60 years old ($p = 0.005$). CAP BOP was given in full dose to patients under age 70 and at two-thirds dose to patients older than 70. The authors acknowledge that this study is peculiar in that the age limit was low, the patients were a subgroup of patients of all ages entered in a clinical trial where participating physicians may have selected patients for enrolment, and early stage patients were included. In addition they state that many of the unrelated deaths were cardiovascular in origin and it is unclear whether these deaths were due to the stress of NHL treatment or the result of underlying cardiovascular disease.[8]

The GELA group carried out the only randomized study treating 273 patients older than 69 years who presented with aggressive lymphoma between October 1987 and September 1989 comparing chemotherapy with cyclophosphamide, teniposide, and prednisolone with (CTVP) or without (CVP) THP-adriamycin (pirarubicin). The median age was 74 years and the overall response to treatment was 47 per cent CR with 34 per cent for CVP and 60 per cent for CTVP ($p \leq 0.001$). Twelve per cent of patients given CVP died during therapy versus 18 per cent treated with CTVP. The median overall survival was 14 months, freedom from progression survival 7 months, and median freedom from relapse survival was not reached with a median follow-up of 8 months. Overall survival was not statistically different in the two treatment arms but CTVP patients had better freedom from progression survival ($p \leq 0.05$) and freedom from relapse survival

($p \leqslant 0.05$). According to the authors, age did not influence the response to treatment, although a better outcome was statistically associated with pirarubicin treatment despite relatively more substantial toxicity.[9]

The Aviano group treated prospectively between January 1987 and April 1990, 52 consecutive patients aged 70 years or older (median age 75 years) with a new combination of etoposide, mitoxantrone, and prednimustine (VMP) specifically devised for elderly patients with NHL. VMP is actually composed of effective and relatively safe single agents and moreover vepeside (VP16), prednimustine and mitoxantrone are part of two active and safe regimens presently used in malignant lymphoma. VP16 and prednimustine in combination with carboxy chloroethyl nitrosourea (CCNU) (the CEP regimen) are used in Hodgkin's disease[10] and prednimustine and mitoxantrone in the NOSTE (Novantrone-Sterecut) regimen in NHL.[11] Among the 48 evaluable patients, the CR rate was 44 per cent while toxicity was mild. The median survival was 12 months. The relationship between relative dose intensity (RDI) and some clinical characteristics of the patients was analysed. Whereas performance status and stage were not related to RDI, the patients who experienced grade IV haematological toxicity received a significantly lower RDI than those without such toxicity. Furthermore it should be noted that the RDI *per se* did not seem to influence the response.[12]

These prospective studies suggest that older patients with aggressive NHL should be treated with curative intent. However, specifically devised regimens for the elderly should be given, because excessive treatment-related deaths could be observed with the conventional chemotherapy regimens. On the other hand, the high frequency of death from causes other than therapy should not stop the treatment of an otherwise rapidly fatal disease. However, on this point, some of the so-called unrelated cardiovascular deaths could actually be related to anthracyclines included in the chemotherapy regimens. It is also very important to look at the selection of patients and at the median age of the patient population included in these trials. A 5 year difference in patient population included in two different trials is much more significant and important in older than in younger patients. For instance, if a patient population of 70 years of age is compared with a population of 75 years of age, the difference in toxicity and CR rate is likely to be higher than if we compare a patient population of 50 years with a population of 55 years of age. Therefore the results of series without a large percentage of older patients (more than 60 years of age) should be taken with caution and it should be clearly stated that these results are valid only for young and selected patient populations with NHL. However, it must be recognized that only from prospective and possibly randomized studies can better knowledge of the optimal treatment of elderly patients with NHL be obtained. Within the European Organization for Research and Treatment of Cancer (EORTC) Lymphoma

Group there is an ongoing trial with more than 50 patients entered so far comparing VMP, previously reported in the Aviano experience, with classic cyclophosphamide, doxorubicin, vincristine, prednisone (CHOP) in diffuse aggressive NHL.

The four previously mentioned groups are actively participating to an international study group of NHL in the elderly constituted during the Venice conference, 15–16 October 1990, with the aim of analysing data both retrospectively and prospectively in a cooperative way. A central data system has recently been activated within the study group.

Toxicity

The toxicity which can be observed in unselected patients older than 70 years of age with NHL treated with conventional chemotherapy regimens is high and probably unacceptable. In a group of unselected patients older than 70 years of age with a diagnosis of diffuse histiocytic lymphoma, Armitage *et al.* reported 30% treatment-related deaths with CHOP combination chemotherapy.[13] The largest study on toxicity associated with chemotherapy in elderly patients with NHL was performed by the EORTC Lymphoma Group.[14] During 1984, 137 cases of elderly NHL were seen at the cooperating institutions, making up 28 per cent of the total number of NHL seen at these institutions.

Most of the physicians used standard therapy regimens at reduced doses from the beginning of the treatment. As expected, severe and lethal toxicity was significantly increased in patients treated with aggressive therapy, defined as polychemotherapy regimens or extended field radiotherapy, in comparison with conservative treatment, defined as one or two antineoplastic drugs or local field radiotherapy.

Among the 123 patients for whom a complete toxicity history was reported, this was severe (grades III and IV according to WHO) in 23 per cent of the patients treated with aggressive regimens compared with 6 per cent of patients treated with a conservative treatment ($p = 0.01$). Moreover, lethal toxicity was not observed among the patients in whom conservative treatment was used, whereas eight of the 71 patients who underwent aggressive treatment died of treatment-related toxicity ($p = 0.02$). Causes of treatment-related deaths included infection in seven and intestinal perforation in one patient. Six of the eight treatment-related deaths occurred in patients who received a standard dosage chemotherapy as used in younger patients.[14] It is also important to be aware that in contrast to younger patients, severe and lethal toxicity in the elderly can be misdiagnosed as cardiac or infectious diseases due to their age, rather than as treatment related.

It is clear that the dose intensity of the chemotherapy given is also very important in elderly patients with NHL in achieving sustained CR and cure. However, these patients were less likely to receive full doses of chemotherapy even during the first cycle of chemotherapy. It is appealing to hypothesize, but is by no means proven, that the inability to deliver full-dose chemotherapy to many elderly patients because of poor treatment tolerance is the reason for the poorer results observed in the group of older patients. Additionally, cutting the starting dose does not improve response in the elderly. The CHOP studies of the Southwest Oncology Group (SWOG) specified a reduction at the beginning in dose by 50 per cent for cyclophosphamide and adriamycin for patients older than 65 years of age, but only a 37 per cent CR rate was observed in this population.[15]

There is a need therefore to look at new combinations that could be given at full-dose regimen but with an acceptable toxicity in order to increase the CR and survival rate in patients older than 70 years of age. In this setting, growth factors could represent the most promising drugs to be tested. Clinical studies are underway in patients older than 70 years.

Future directions

Elderly patients should be included in clinical trials of NHL with the aim of developing specifically devised chemotherapy regimens, possibly in conjunction with new therapeutic modalities such as growth factors. Quality of life of elderly patients with NHL has to be measured carefully and possibly improved.

Moreover, the following issues should be taken into consideration for the improvement of knowledge of the natural history and for the optimal treatment of elderly patients with NHL.

1. New regimens or modalities of treatment employed in elderly patients with NHL should be evaluated carefully. In particular it must be underlined that the results obtained are valid for a population of patients with that median age, and inclusion criteria for these studies should be reported clearly. Moreover, careful follow-up of elderly patients treated with a new regimen should be carried out in order to find possible severe and lethal toxicity associated with that treatment which may not be evident immediately.
2. The clinical pharmacology of antineoplastic agents should be studied in patients older than 70 years of age. Combination chemotherapy regimens now employed in elderly patients with NHL are based on single-agent pharmacokinetics data collected in patients usually of less than 50 years of age. Furthermore, phase II studies should be carried out in elderly patients.
3. In order to increase the dosage of chemotherapy actually given and

possibly improve the subsequent CR rate and overall survival, growth factors should be studied fully in this age population, trying to overcome the bone marrow toxicity that usually limits chemotherapy in elderly patients. Many studies are underway in this area and significant results may be achieved in the next few years.

Key points

- Approximately one-third of patients with non-Hodgkins lymphoma are aged 70 years or older.
- Histological grade is not affected by age but extranodal disease is more common in elderly patients.
- The few prospective studies suggest that older patients should be treated with curative intent.
- Because of increased toxicity in the elderly, dose modification with growth factors to reduce side-effects requires testing in this age group.

References

1. National Cancer Institute Update. (1986). *Annual cancer statistics update, December 8, 1986*. National Cancer Institute, Bethesda, MD.
2. Exton-Smith, A.N. (1982). Epidemiological studies in the elderly: methodological considerations. *American Journal of Clinical Nutrition*, **35**, 1273–9.
3. Kennedy, B.J. (1988). Aging and cancer. *Journal of Clinical Oncology*, **6**, 1903–11.
4. Zagonel, V., Carbone, A., Kerper-Fronius, S., Kuse, R., Jelic, S., Huber, H., Tirelli, U., Ludwig, H., Ptesnica, S., and Monfardini, S. (1990). Management of non-Hodgkin's lymphomas in the elderly patients: conclusions of the Second Intercity Meeting of the European School of Oncology (ESO), 1987. In *The management of non-Hodgkin's lymphomas in Europe*, (ed. S. Monfardini), pp. 35–44. Springer Berlin.
5. Carbone, A., Volpe, R., Gloghini, A., Trovo, M., Zagonel, V., Tirelli, U., and Monfardini, S. (1990). Non-Hodgkin's lymphoma in the elderly: 1, pathologic features at presentation. *Cancer*, **66**, 1991–4.
6. Tirelli, U., Zagonel, V., and Monfardini, S. (1991). Common errors in conducting and reporting clinical trials in non-Hodgkin's lymphomas and patients' age. *European Journal of Cancer*, **6**, 811.
7. O'Reilly, S., Klimo, P., and Connors, J.M. (1991). LD-ACOP-B and VABE: weekly chemotherapy for elderly patients with advanced stage diffuse large cell lymphoma. *Journal of Clinical Oncology*. **9** (5), 741–7.
8. Vose, J.M., Armitage, J.O., Weisenburger, D.D., Bierman, P.J., Sorensen, S., Hutchins, M., Moravec, D.E., Howe, D., Dowling, M.D., and Mailliard, J. (1988). The importance of age in survival of patients treated with chemotherapy for aggressive non-Hodgkin's lymphoma. *Journal of Clinical Oncology*, **6**, 1838–44.

9. Coiffier, B., Gisselbrecht, C., Bosly, A., *et al.* (1990). Treatment of aggressive lymphomas in patients older than 69 years. First interim report of a randomised study from the GELA. *Fourth international conference on malignant lymphoma, Lugano, 6–9 June*, Abstract No. 82.

10. Bonadonna, G., Viviani, S., Valagussa, P., Bonfante, V., and Santoro, A. (1985). Third line salvage chemotherapy in Hodgkin's disease. *Seminars in Oncology*, **1**, 23.

11. Landys, K. and Ridell, B. (1986). Novantrone in combination with sterecyt in treatment of unfavourable non-Hodgkin's lymphoma. *Fifth NCI–EORTC symposium on new drugs in cancer Therapy, Amsterdam*, Vol. 22, p. 9.43.

12. Tirelli, U., Zagonel, V., Errante, D., *et al.* (1991). A prospective study on a new combination chemotherapy with etoposide, mitoxantrone and prednimustine in 52 patients aged 70 years or older with unfavourable non-Hodgkin's lymphoma. (Submitted for publication.)

13. Armitage, J.O. and Potter, J.F. (1984). Aggressive chemotherapy for diffuse histiocytic lymphoma in the elderly: increased complications in advancing age. *Journal of the American Geriatric Society*, **32**, 269–73.

14. Tirelli, U., Zagonel, V., Serraino, D., Thomas, J., Hoerni, B., Tangbury, A., Ruhl, U., Bey, P., Tubiana, N., and Breed, W.P. (1988). Non-Hodgkin's lymphomas in 137 patients aged 70 years or older : a retrospective European Organization for Research and Treatment of Cancer Lymphoma Group Study. *Journal of Clinical Oncology*, **6**, 1708–13.

15. Dixon, D.O., Neilan, B., Jones, S.E., Lipschitz, D.A., Miller, T.P., Grozea, P., and Wilson, H.E. (1986). Effect of age on therapeutic outcome in advanced diffuse histiocytic lymphoma: the Southwest Oncology Group experience. *Journal of Clinical Oncology*, **4**, 295–305.

13

Palliative and terminal care

S. Ahmedzai

Interface between oncology and palliative care

The term 'palliative care' is increasingly replacing 'terminal care' for the management of patients with advanced disease for whom curative treatment is not appropriate, and who may be approaching death. The precise cut-off points for when a patient has 'advanced' disease or is near to death, are not of prime importance here (although they are significant for those who are planning the provision of care to this group).[1,2]

'Palliative care' and 'palliation' are derived from the latin word 'pallium', which meant a soldier's cloak.[3] From the traditional layperson's meaning of to 'palliate' in the sense of 'covering up' problems (sometimes with the negative connotation of 'hiding'), the medical usage has acquired a more positive meaning of actively relieving problems. In many cases it is not clear whether it is the symptoms of a disease (such as pain from bony metastases), or the underlying disease process itself (for example progressive prostate cancer) which is being palliated. Indeed, in many instances palliative care has to be directed at both symptoms and underlying disease to be effective and worthwhile.

Oncologists have long defined 'palliative' intent of treatment, either in the context of an individual's treatment plan or in respect of trial protocols, as being distinct from 'radical', 'curative', or 'adjuvant'.[4,5] For a long time there has also been an underlying assumption that radical or curative meant 'active', while palliative was 'passive', or in some other way had a less worthy connotation. Thus while there has been a call for increased 'flexibility' in the planning of radiotherapy for older patients,[6] it is also claimed that adapting curative treatments in old age to reduce toxicity would be 'compromising the chance of cure'.[7] It is perhaps more productive to view the differences between palliation and cure broadly in terms of the realistic prospects of useful prolongation of life (which would need to be 'age-adjusted' for the patient group under consideration), off-setting this against the anticipated toxicities and net effect on quality of life. There is a case for recognizing a third distinct group of treatment intentions: 'terminal' care, where the patient is expected to die very soon and there is no attempt to prolong life, and all toxicity is to be avoided.[8]

Physicians need to be aware that their perceptions of toxicity may be unduly pessimistic and it is possible that they are also somewhat 'ageist' in attitude. Where controlled studies have been carried out, it can be seen that the classical adverse effects of 'aggressive' treatment may be well tolerated by elderly patients. Good examples of this have been seen in chemotherapy for small-cell lung cancer[9] and surgery for colorectal cancer.[10] Although the latter study concentrated on surgical complications, another surgical series for benign oesophageal stricture which used a patient-oriented subjective evaluation showed clear benefits to quality of life.[11]

The distinction between palliative care and treatment with curative intent is often particularly difficult to discern in the case of the elderly cancer population. In spite of increasing longevity in the general population and the improved success of cancer treatments, the benefits in terms of increased 'long-term' survival may be less apparent to older than to younger patients. This is not just because the classical notions of 5 or 10 year survival are inappropriate when the expected length of life may in any case be in single figures.[12,13] The benefits of living longer after cancer treatment are also dubious for a group which is susceptible to multiple pathology, especially from degenerative diseases, so that cancer-related morbidity or mortality may be less significant than that from other concomitant diseases.[14] Furthermore, it is recognized that elderly patients are prone to suffer from a 'cascade' of problems, often arising from inadvertent adverse effects from inappropriate therapies, or even the act of hospitalization.[15]

These observations should not, however, lead oncologists to adopt a negative attitude towards treating older patients with advanced cancer. With careful selection of therapy based on individual needs, even relatively modest prolongation of life, if it is accompanied by symptom control and tolerable toxicity, may be very useful to some. It is clear that the perspective of clinicians may be at variance with their patients' own aspirations, so that seemingly small survival benefits from potentially toxic treatment which would be rejected by doctors, could be welcomed by patients themselves who have much more at stake.[16] This study did not differentiate patients' responses by age group, and it is interesting to speculate whether the desire for life is as strong with elderly cancer sufferers, when faced with these choices.

Holistic approach to palliative care

In the new medical specialty of 'palliative medicine', the basic philosophy is the need to adopt a holistic and multidisciplinary approach to the care of patients.[17] This simply means regarding the individual as a whole person, and broadly it encompasses the physical, psychological, social, and spiritual domains of life. It requires a constant balancing act to keep these different but overlapping needs met as far as is possible, within the constraints of the patient's desires and local resources. The latter includes the coordinated use of all relevant specialities—medicine, nursing, physiotherapy and occupational therapy, dieticians, trained counsellors, social workers, and chaplains.

In palliative care the needs and problems of carers—both informal (family) and formal (professionals or voluntary)—are also addressed. Another sensitive issue which is naturally linked to the subject of relief of terminal suffering is euthanasia, but this complex area is beyond the scope of this chapter.[18] Finally, palliative care of cancer patients would be incomplete without some attention to the problems of bereavement in relatives. However, this chapter will concentrate on the patient's issues, except where the family context is highly relevant, as with social or sexual problems.

Physical aspects

The physical aspects are the easiest of the four broad domains of holistic palliative care to approach, and will be examined here as a model for the others. Most of the research in palliative care provision has centred on the relief of physical problems.[19] However, it is clear from reviewing the literature that the specific problems of elderly patients have not usually been studied separately in a controlled, prospective way. As palliative care provision so often focuses on the older age groups in society, evidence of their special needs may sometimes be extrapolated from the existing general palliative care literature. However, this retrospective approach is not a sound basis for determining the true prevalence or relief of problems, as a retrospective survey of deaths from all causes in the UK has shown.[20] Even with a preponderance of patients in this study over the age of 65, those who had died from cancer turned out to be younger than non-cancer patients. The cancer patients' symptoms and restrictions on life before death were more severe but of shorter duration. Simple assumptions about elderly cancer patients based on general demographic data for elderly people may therefore be suspect.

It would be very helpful for studies of symptom control in cancer patients across the age spectrum, to present the findings broken down by different age bands. Even negative results—such as the observation that older patients' problems and results were no different from those of younger patients—would be positively helpful in dispelling some of the myths of 'ageism'. The major methodological issues of relevance here, and also with reference to the non-physical aspects, are how to measure physical symptoms, who should measure symptoms, and what criteria should be used to evaluate the success of treatment?

How to measure symptoms?

The measurement of physical distress in cancer patients has been extensively researched and reviewed.[21] In oncological practice, as elsewhere in medicine, it is important to use instruments that are reliable and valid, but at the same time are feasible in the clinical situation. The major division lies between the use of linear/visual analogue scales (often abbreviated to LAS or VAS), or verbal rating or categorical scales (VRS or CS). Each of these approaches has its benefits and they have been shown to correlate well when used in the same patients. In general, the experience in the UK is that with very sick or elderly patients, it is easier to explain verbal scales rather than the more abstract analogue scales. This has not been formally tested, and neither has the issue of compliance with verbal versus analogue scales in repeated measures, in older patients.

A third method, favoured by the Medical Research Council in the UK, is the patient's daily diary card.[22] This has been shown to be a sensitive way of recording toxicities and broad aspects of 'quality of life', but has not been extensively used in studies of specific physical symptoms. An a priori judgement (to be confirmed) is that older patients may also find daily diary cards more difficult to comply with over long periods.

It is not appropriate here to go into detail about the construction and psychometric validation of subjective scales; this is covered in Chapter 10, and by other reviews published elsewhere.[23,24] Wherever possible, the clinician is urged to use an existing validated instrument. If one does not exist to cover the particular symptoms he is interested in, for example the problems of cutaneous fungating tumours, it is best to design and pretest new questions in collaboration with psychologists and statisticians used to this field.

It is worth mentioning here the practical approach of the European Organization for Research and Treatment of Cancer (EORTC) modular questionnaire on quality of life.[25] This simple concept is to use a 'core questionnaire' which covers those areas—physical, psychological, social – that affect most cancer patients, and then to attach to this one or more extra modules each consisting of a few items relating to a special clinical topic. The standardized core questions and extra modules should make for easier comparison between studies, and could make extraction of data on specific age groups and subsequent meta-analysis again easier and more reliable. The EORTC instrument, which is composed of 'Yes/No' and four-point categorical scales and also has hybrid numerical/verbal scales on global 'physical condition' and 'quality of life', has the extra advantage of having been translated and validated in many European languages.[26]

It is not yet possible to identify well-conducted studies in which physical symptoms have been specifically addressed in defined older populations. Studies on cancers that are more prevalent in the elderly may, by inference, be used to shed light on the symptoms of this age group. Thus in a trial of estramustine versus mitomycin for patients with prostate cancer, symptoms were measured by means of patient- and physician-completed evaluations.[27] Unfortunately, no breakdown of the data by age is supplied. However, it may be reasonably assumed that most patients were middle-aged or elderly. The interesting finding was that doctors tended to underestimate pain, nausea, and decreased performance status compared with the patients' own assessments. In spite of the considerable practical problems of giving out and collecting patient questionnaires, which led in this study to very poor overall compliance (itself a possible source of bias in the results), the authors concluded that 'monitoring of subjective morbidity by the patient should be mandatory in cancer trials where palliation is a major endpoint'. It seems quite reasonable to argue further, that where palliative care or terminal care is the primary therapeutic objective, patient-rated evaluations should of course be mandatory.

Who should measure symptoms?

It would be wrong to assume from the previous paragraph that physicians' (or nurses') assessments are of no value. There are situations that are familiar to all sensitive clinicians where patients may not be the best or most reliable 'witness'.[28] Some patients, particular elderly patients, may be unable to complete evaluations because of confusion, dysphasia, or dementia. It would be unreasonable and biased to exclude physical distress of such patients in any formal evaluation, and so proxy measures must be devised. Simple, standardized, and pretested verbal rating scales which are completed by the clinician are probably the best solution here, although questioning of relatives, with careful interpretation, may be helpful.[20,29]

Another source of bias in the measurement of physical symptoms in palliative care is the failure to take into account the natural changes in physical parameters or performance due to age itself. An example of this is the progressive loss of lung volumes and pulmonary function with advancing years, which could confound a study on respiratory problems in lung cancer.[30] The observation that asthma is often under-diagnosed in older patients could also lead to an overestimate of dyspnoea from assumed pulmonary involvement from malignancy.[31]

Weakness is so often a feature of advanced cancer that it is usually taken for granted, and increased severity is also assumed to be causally related to the progressive malignancy. Elderly patients especially are prone to many biochemical, haematological, and skeletal conditions which predispose to weakness, fatigue, and reduced physical performance.[32] A good example is the common usage of diuretics in the elderly, which may lead to hyponatraemia or hypokalaemia, either of which could cause or aggravate weakness. Sodium and water-load handling may be impaired in elderly subjects, perhaps related to changes in antidiuretic hormone (ADH) homeostasis.[33] The effects of hyponatraemia from impaired water handling include anorexia, confusion, headaches, cramps, nausea, and vomiting. Since many of these occur as physical symptoms of malignancy itself, the possibility arises of misinterpreting the results.

There is a danger of failing to allow for concomitant medication in interpreting findings on physical symptoms, as exemplified by the hypnotic drugs. Sleep disturbance is common in the elderly, and one-third of all prescriptions for hypnotics are for patients aged over 65 years.[34] In palliative care great attention is paid to the quality of sleep, and apparent benefits in sleep patterns from improved pain control or reduction in cough, may in fact have other origins. It is in fact very difficult in many clinical situations to extract which intervention was responsible for a particular effect. In trial protocols, therefore, it is important to restrict the use of other drugs as much as possible, or if that is not considered ethical, to record these in detail.

It is often stated anecdotally that older patients have different thresholds for some symptoms, for example pain. At least with post-operative pain, there is good evidence that older patients require less analgesic medication for the same degree of pain control.[34] In less controlled populations there are many issues which could lead to a biased interpretation of symptom threshold: failure to record other concomitant medication (especially sedatives and other psychoactive drugs), fear of undue adverse effects (sometimes justified), failure to allow in heterogeneous cancer samples for differential predisposition to painful metastases, different expectations and opportunities for physical activity, and not least the inherent bias of clinicians who believe that older patients need less analgesia. Even if a true difference in palliative drug requirement were to be found, the mechanism for this needs to be clarified. Could it be reduced metabolism or excretion of the drug? Or is there a genuine shift in physical perception of pain with age?

How are the criteria for successful treatment in palliative care determined?

This question is central to the application of palliative care itself (or of a palliative regime in a cancer trial). If relief of a symptom cannot be demonstrated, three possible interpretations exist: the intervention was actually unsuccessful, the instrument or criteria for evaluating it were insensitive or inappropriate, or, most difficult, other confounding factors make the resolution of one symptom impossible to identify. The first interpretation is straightforward, and clearly one which a trial should be designed to demonstrate or refute. Even then, there is a grey area of debate about how much improvement constitutes 'relief' of a problem. An apparent advantage of analogue scales is that minute differences can be measured (down to 1 mm changes over a 100 mm line). Thus statistically 'significant' differences may sometimes be shown after a treatment, but the question arises of the 'biological' significance of the few millimetres shift. A further problem to compound this is the use, inappropriately, of parametric statistics such as Student's t-test, when the requirements for normality and sampling have not been met.

It is helpful to designate, before analysis, how much of a shift along the analogue scale represents a clinically useful response. This may involve piloting the scale with additional, patient-rated measures for external validation. Alternatively, sections of the 100 mm line may be subdivided into regions representing 'mild', 'moderate', and 'severe' symptoms. This would rather tend to negate the rationale of using an analogue scale in the first place. Even with verbal scales, however, it is helpful if the protocol or analysis plan states how much shift represents a clinically significant response.

A negative result may be due to the fact that the statistical test used was insensitive to the data being analysed. Expressing this more broadly, the instrument itself may have been insensitive or inappropriate, for example the excessive reliance on patient-completed forms in a group predisposed to confusion or dysphasia.

If a study in palliative care attempts to measure the changes in one symptom without taking into account possible influences from other problems, there is a risk of the results being inconclusive. With elderly cancer patients it is quite likely that many other factors, including psychosocial and cultural ones, may mask the true benefit brought about by a treatment. For example, a clear improvement after a new analgesic treatment may be recorded in pain level, but owing to other aspects of multiple pathology such as immobility due to weakness, bone metastases, or paraplegia, the patient does not become any more mobile. Measuring only the pain would show a 'success', while physical functioning status or other more global aspects of quality of life may not change at all.

Some researchers like to use the consumption of analgesics as a surrogate measure of overall pain control. Indeed, this may be incorporated into a 'formula' for expressing pain relief, by integrating patients' subjective experience, observed behaviour, and analgesic consumption, as was used in a recent study evaluating the benefit of strontium-89 in metastatic prostate cancer.[36] At first sight this appears to be practical (it is easy to record drug consumption) and clinically useful (reduced drug consumption may mean improved symptoms and/or reduced adverse effects). However, both contentions merit scrutiny. Patients' own recall of drug consumption is notoriously unreliable, unless very strict guidelines are followed or the patient is hospitalized. Secondly, in good palliative care practice adverse effects are usually controllable, and 'significant' reductions in opioid intake after new interventions such as radiotherapy are often not forthcoming. Indeed, in hospice care it is usually seen as a sign of better control that the patient learns to take more rather than less analgesia, on a regular rather than 'as required' basis. Clearly, however, if a new treatment reduces the need for other major interventions such as courses of further radiotherapy or episodes of hospital admission, the social and economic benefits may be considerable, especially for elderly patients living at home.

Psychological aspects

There is a great deal of information about the attitudes and psychological responses to advancing ill-health and approaching death in the 'normal' population. For many old people, increasing infirmity and loss of bodily functions and control may lead to a more ready acceptance of death, than for younger patients.[37] More and more elderly people find themselves in positions of receiving 'care'—in long-stay wards of hospitals, in residential care or private nursing homes, sometimes in hospices. Institutionalized care does not automatically lead to loss of independence or dignity, but may predispose to it. The loss of loved ones of the same generation, or even of familiar pets, may add to the isolation. Many such patients may even welcome death, especially if life is accompanied by unrelieved physical suffering. Interestingly, in the retrospective UK survey it was found that older cancer patients were more likely to have relatives or other companions at home than non-cancer patients of the same age group.[20]

However, often there is a psychological reaction by the older person to his or her predicament. The psychological aspects of cancer itself have been discussed in Chapter 10 and elsewhere.[8,39] The emotional impact on carers of cancer patients has also been well documented, albeit in a heterogeneous group spanning 18–79 years.[40]

Mild anxiety is a common, almost universal, reaction to learning the diagnosis of cancer. This may become a chronic state or may lead to frank psychiatric disturbance. It is probable that older patients are less liable to this than younger patients.[39] Benzodiazepines are frequently prescribed for anxiety and insomnia in the elderly, although it is thought that the ageing brain is more susceptible to this class of drugs, and their metabolism may be impaired by the process of ageing or other age-related conditions.[41]

Another common reaction to cancer is depression, which is inherently more prevalent in the over 60 age group of the general population,[42] and especially among those resident in nursing homes.[43] Its diagnosis is made difficult in elderly cancer patients by the simultaneous presence of many of the classical symptoms of depression—anorexia, lethargy, poor sleep—which may in fact be caused by the malignancy. Full evaluation may need a formal psychiatric assessment, which is difficult in many clinical trial settings and may be inappropriate for elderly patients living at home. Treatment is also fraught with problems, as drugs are slow in action and may have significant adverse effects, especially in the elderly, for example cardiac arrhthymias or retention of urine. Again, no formal study has tried to evaluate the role of antidepressant treatment specifically in elderly cancer patients.

A review of cancer patients referred to an American psychiatric service showed that, in contrast to the general population, depression was more likely to occur below the age of 60, while organic brain syndrome (confusional state) was seen more in those over 60 years old.[44] This has led to the warning that while general psychosocial adaptation to cancer may be maintained into old age, the increased risk of dementia in elderly cancer patients raises problems regarding informed consent for treatments.[45]

In palliative care there is a great emphasis on effective communication between the patient and the professional team. Trained staff are often available in hospices or in the patients' homes to counsel.[2] Newer behavioural approaches to help patients control the psychological stresses of cancer are being incorporated into oncological clinics. Until now, trials of these interventions have not controlled for age, and it is possible that older patients may be inadvertently or deliberately excluded from these therapies. In a prospective American study of group therapy for breast cancer patients the average age was 54.7 years,[46] and in a randomized Scottish trial of relaxation training, the mean ages for men and women were 39.7 and 48.5 years respectively.[47]

A particularly sensitive psychological aspect in cancer treatment is the effect on sexual desire and function. Many malignancies directly or indirectly affect sexual organs, or the treatments may involve hormonal manipulations that may reduce libido, such as the antiandrogen therapies in prostate cancer. The effect of surgery on body image, especially in the case of mastectomy or pelvic operations, can have very damaging psychological consequences. It is now recognized that sexual function is still important in older people, even in couples living in residential accommodation.[48] There has been relatively little regard for this area in cancer studies up till now. It is important for patients in prostate cancer trials, and older patient groups with other sexually oriented cancers, to be asked about their sexual feelings and function. This will have to be done sensitively, and once again the use of standardized questionnaires may be more reliable (and less embarrassing) than *ad hoc* questions. Ideally the views of sexual partners should also be considered. The recognition of sexual problems should lead to the offering of treatment. It would appear that in medical practice there are a number of psychological barriers to be overcome on the side of the professionals, as well as the patients, before this becomes routine.[49]

Reference has already been made to the disturbance of body image resulting from some cancers or treatments. It is not known whether older patients suffer more or less in this respect, although some studies have implied that the elderly may accept more readily the physical and psychosocial changes caused by treatment.[38] The employment of specialist nurses, such as for stoma care or after mastectomy, has increased recently, but formal evaluation of these expensive posts is required.

Social aspects

Physicians treating cancer patients have long recognized the importance of social factors in illness and management. Social circumstances often determine the need for hospitalization, and may influence significantly the length of stay and readmission rates, as much as physical aspects of the disease process. In acknowledging the social aspects, we must also give credit to the role of informal carers (usually family members) in supporting the patient.[50]

Much of the earlier work on the problems faced in terminal disease was based on interviews with relatives, and included good coverage of the social deprivations associated with dying.[51] Studies on the provision of domiciliary palliative care by multidisciplinary teams have also included evaluation of social problems.[52] It is difficult to identify socially oriented research conducted specifically on elderly cancer patients, or even reports in heterogeneous cancer samples where sufficient numbers have allowed for the analysis of results by age groups.

In measuring social effects, the use of scales or profiles for social functioning and activities of daily living (ADL) are useful.[53] The Katz ADL scale is a standard instrument, and was used in a prospective study of terminal cancer patients being followed up at home in Sweden.[54] It was found that age, sex, or type of cancer did not influence the ADL status, but the latter did correlate with need for assistance and length of care required. It would thus appear that the ADL scale could be used in future studies of elderly cancer patients without being subject to bias by age group itself.

It is often stated that 'home is the best place' for the cancer patient. This would appear to be true also for most elderly people. However, considerations of financial hardship, which in a cold climate may predispose old people who are living alone to hypothermia, should be taken into consideration.[55] In rural areas, older patients might be expected to prefer to remain at home: in one study in the Lake District of England, those under 70 were more likely to die at home, while more over this age died in hospital.[56] In contrast, a review of deaths in the Oxford area of England showed that people aged 85 and over were less likely than the 65–84 age group to have been admitted to hospital in the last year of life.[57] Once admitted, however, the older patients spent more time in hospital. It is a pity this report did not differentiate between diagnostic groups, as the large number (over 140 000) would have made such an analysis feasible.

The choice regarding place of death was reviewed by a London hospital palliative care support team, and although 53 per cent would have preferred to die at home, 63 per cent actually died in hospital.[58] There was no age effect on place of death. In another prospective study 58 per cent of cancer patients expressed the wish to die at home, but this figure rose to 67 per cent if the home circumstances had been more favourable.[59] How to make home circumstances more favourable for terminal care needs to be the subject of more controlled and randomized trials, but the difficulties and ethics of setting these up are formidable. The specific social issues that have to be resolved in the provision of terminal care include financial support, relief for informal carers (for example by sitting services, day care for patients), aids for living through occupational therapy, ancillary services such as laundry and home meals. Most of these are common to patients dying in all age groups, but older people are more likely to be living alone, or to have a carer or a dependent who is equally aged and infirm.

Spiritual aspects

By 'spiritual' I mean not only the religious needs of the patient, but also the existential aspects—the soul-searching questions that arise in response to the diagnosis of cancer, such as 'Why me?', 'Why now?'[60] In older people with naïve or prejudiced views about the aetiology of cancer, there may be deep shame about having the disease—it may also be seen as divine punishment for earlier (often minor) wrong-doing. Such thoughts are very private, and will usually only be shared with close family members or trusted professionals, such as in a hospice setting.

Should such themes be opened up at all? There is no 'hard' evidence that exploration and subsequent counselling for inappropriate guilt or shame, or a sense of personal failure, is necessary or beneficial. On humanitarian grounds, however, such an approach is deemed worthy in palliative care services. The attachment of a social worker, nurse, or chaplain trained in communication and counselling skills may be essential to ensure the 'holistic' process of palliative care.[61] The evaluation of such placements is difficult to formulate using traditional measures of success. The psychosocial scales referred to above may give only crude proxy measures of a patient's spiritual health. Rating scales of spiritual needs are rare, and more testing needs to be done in a variety of clinical settings.[62] To measure truly existential as opposed to religious distress, an instrument needs to be free of religious terminology or implied dogma.

Cultural problems

Some patients may choose to replace conventional medicine or to supplement it with non-orthodox treatments. In one postal survey, 15 per cent of those attending non-orthodox practitioners were over 65 years old.[63] Most of these had musculoskeletal disorders. The true incidence of cancer patients seeking alternatives to conventional oncological therapy is not known, and probably varies according to country and culture.

The cultural and religious differences between cancer patients across the world is a fascinating though difficult subject for research. Even the use or avoidance of words such as 'cancer' and the readiness of doctors to discuss death and dying may have an impact on the subjective experience of patients. It is easy to assume that the western, Christian-based approach to open acceptance of death from advancing cancer is universal or indeed desirable. The Jewish emphasis on physical life on this earth leads to the opposite view, namely always to hold on to hope for continuing life.[64]

One obstacle to cross-cultural studies is the problem of translating standardized instruments. The EORTC quality of life questionnaire, by being available in most European languages, clearly has an advantage here.[26]

Priorities for palliative care in elderly cancer patients

A brief but moving statement about how it felt for a doctor to watch her father die in hospital from laryngeal cancer, recently shocked many readers of the *British Medical Journal*.[65] Even the title of the anonymous piece, 'Terminal careless', seemed calculated to remind the medical profession that it still had a long way to go in learning how to apply what is already known about good palliative care into everyday practice.

There is evidence that patients even in the last week of life from breast cancer are being heavily investigated and treated with chemotherapy.[66] While these interventions may be appropriate for some, for elderly patients they are usually unnecessary and distressing. Or are they? Can we reasonably make assumptions from our inbuilt prejudices and biases, sometimes based on a single personal experience, which could influence how we and others should practice palliative care for the elderly?

One way to prevent idiosyncratic and prejudiced treatment is for clinicians caring for the elderly to work in a multidisciplinary framework. This is already the accepted *modus operandi* in palliative care: it should be adopted in wider areas of oncology. As I have indicated many times above, however, I would not expect any such change in care to come about without some attempts at formal evaluation. Teams should be audited, just as we accept for individual treatments.[52] It is healthy to be sceptical about multidisciplinary working, if only to prevent a new dogma from replacing the 'dreary dictatorial days of consultant (specialist physician) power'.[67]

In the conclusions to the report of the Joint National Cancer Institute (NCI)–EORTC Consensus Meeting on Neoplasia in the Elderly, it was rightly stressed that older cancer patients should be treated like any younger patient, without prejudice based on age.[68] I also support the statement that specific psychosocial protocols should be designed and adopted. I would add to the recommendations that wherever possible, palliative care physicians, nurses, and other professionals should be included in the multidisciplinary team caring for the elderly patients and their families.

Further reading

Doyle, D., Hanks, G.W.C., and MacDonald, N. (eds.) (1993). *Oxford textbook of palliative medicine*. Oxford University Press.

Key points

- Elderly patients may be less affected by side-effects than is preconceived by their doctors.
- Verbal rather than analogue scales may be better for subjective assessment of symptoms in the elderly.
- Social functioning can be measured by the Katz activities of daily living (ADL), which is unaffected by the age of the patient.
- Although more than 50 per cent of patients with terminal cancer will wish to die at home, circumstances may be unfavourable in the elderly.
- It is important that those caring for the elderly with advanced cancer work within a multidisciplinary framework.

References

1. Mor, V. and Masterson-Allen, S. (1990). A comparison of hospice vs conventional care of the terminally ill cancer patient. *Oncology*, **4**, 85–96.

2. Ahmedzai, S. and Wilkes, E. (1988). Dying with dignity : a British view. In *Cost versus benefit in cancer care*, (ed. B.A. Stoll), pp. 73–80. Macmillan, London.
3. Saunders, C. (1987). What's in a name? *Palliative Medicine*, 1, 57–61.
4. Maher, E.J., Dische, S., Grosch, E., Fermont, D., Ashford, R., Saunders, M., *et al.* (1990). Who gets radiotherapy? *Health Trends*, 22, 78–83.
5. Lewis, C.R., Kaye, S.B., and Calman, K.C. (1990). Principles of chemotherapy. In *Cancer in the elderly*, (ed. F.I. Caird and T.B. Brewin,) pp. 49–56. Wright, London.
6. Brewin, T.B. (1990). Principles of radiotherapy. In *Cancer in the elderly*, (ed. F.I. Caird and T.B. Brewin), pp. 40–8. Wright, London.
7. Scalliet, P. (1991). Radiotherapy in the elderly. *European Journal of Cancer*, 27, 3–5.
8. Ashby, M. and Stoffell, B. (1991). Therapeutic ratio and defined phases: proposal of ethical framework for palliative care. *British Medical Journal*, 302, 1322–4.
9. Carney, D.N., Grogan, L., Smit, E.F., Harford, P., Berendsen, H.H., and Postmus, P.E. (1990). Single-agent oral etoposide for elderly small cell lung cancer patients. *Seminars in Oncology*, 17 (suppl. 2), 49–53.
10. Lewis, A.A.M. and Khoury, G.A. (1988). Resection for colorectal cancer in the very old: are the risks too high? *British Medical Journal*, 296, 459–61.
11. Jayawardhana, B.N.M., Moghissi, K., and Knox, J. (1989). Quality of life of elderly people after surgery for benign oesophageal stricture. *British Medical Journal*, 299, 1503–4.
12. Lamont, D.W., Gillis, C.R., and Caird, F.I. (1990). Epidemiology of cancer in the elderly. In *Cancer in the elderly*, ed. F.I. Caird and T.B. Brewin pp. 9–15. Wright, London.
13. Kearsley, J.H. (1989). Compromising between quantity and quality of life. In *Ethical dilemmas in cancer care*, (ed. B.A. Stoll), pp. 39–49. Macmillan, Basingstoke.
14. Sclare, G. (1991). Malignancy in nonagenarians. *Scottish Medical Journal*, 36, 12–15.
15. Irvine, R.E. (1990). Sickness and health in old age. In *Cancer in the elderly*, (ed. F.I. Caird and T.B. Brewin), pp. 3–8. Wright, London.
16. Slevin, M.L., Stubbs, L., Plant, J.H., Wilson, P., Gregory, W.M., Armes, P., and Downer, S. (1990). Attitudes to chemotherapy: comparing views of patients with cancer with those of doctors, nurses, and general public. *British Medical Journal*, 300, 1458–60.
17. Tate, T. (1990). Palliative medicine: its content and training. *British Journal of Hospice Medicine*, 44, 140–1.
18. Spreeuwenberg, C. (1989). The right to die. In *Ethical dilemmas in cancer care*, (ed. B.A. Stoll), pp. 63–7 Macmillan, Basingstoke.
19. Ahmedzai, S. (1991). Quality of life research in the European palliative care setting. In *Effect of cancer on quality of life*, (ed. D. Osobal, pp. 323–31. CRC Press, Boston, MA.
20. Seale, C. (1991). Death from cancer and death from other causes : the relevance of the hospice approach. *Palliative Medicine*, 5, 12–19.
21. de Haes, J.C.J.M., van Knippenberg, F.C.E., and Neijt, J.P. (1990). Measuring psychological and physical distress in cancer patients: structure and application of the Rotterdam Symptom Checklist. *British Journal of Cancer*, 62, 1034–8.

22. Fayers, P.M. and Jones, D.R. (1983). Measuring and analysing quality of life in cancer clinical trials: a review. *Statistics in Medicine*, **2**, 429–46.
23. Cella, D. and Tulsky, D.S. (1990). Measuring quality of life today: methodological aspects. *Oncology*, **4**, 29–38.
24. Ahmedzai, S. (1990). Palliative care in oncology: making quality the endpoint. *Annals of Oncology*, **1**, 396–8.
25. Aaronson, N.K., Bullinger, M., and Ahmedzai, S. (1988). A modular approach to quality of life assessment in cancer clinical trials. *Cancer Research*, **111**, 231–48.
26. Aaronson, N.K., Ahmedzai, S., Bergman, B., Bullinger, M., Cull, A., Duez, N.J., *et al.* (1993). The European Organisation for Research and Treatment of Cancer QLC–C30: A quality-of-life instrument for use in international clinical trials in oncology. *J. Natl. Cancer Inst.*, **85**, 365–76.
27. Fossa, S.D., Aaronson, N.K., Newling, D., van Cangh, P.J., Denis, L., Kurth, K.H., and de Pauw, M. (1990). Quality of life and treatment of hormone resistant metastatic prostatic cancer. *European Journal of Cancer*, **26**, 1133–6.
28. Ahmedzai, S. (1990). Measuring quality of life in hospice care. *Oncology*, **4**, 115–19.
29. Ahmedzai, S., Morton, A., Reid, J.T., and Stevenson, R.D. (1986). Quality of death from lung cancer: patients' reports and relatives' retrospective opinions. In *Psychosocial Oncology*, (ed. M. Watson, S. Greer, and C. Thomas) pp. 187–92. Pergamon, Oxford.
30. Sykes, D.A., Mohanaruban, K., Finucane, P., and Sastry, B.S.D. (1989). Assessment of the elderly with respiratory disease. *Geriatric Medicine*, **19** (12), 49–54.
31. Banerjee, D.K., Lee, G.S., Malik, S.K., and Daly. S. (1987). Underdiagnosis of asthma in the elderly. *British Journal of Diseases of the Chest*, **81**, 23–9.
32. Welsh, J. (1990). Terminal care. In; *Cancer in the elderly*, (ed. F.I. Caird and T.B. Brewin), pp. 57–64. Wright, London.
33. Miller, M. (1989). Fluid and electrolyte balance in the elderly. *Geriatric Medicine*, 73–82.
34. Damle, A. (1989). Psychiatric aspects of sleep disorders in the elderly. *Geriatric Medicine*, 67–70.
35. Bellville, J., Forrest, W.H., Miller, E., and Brown, B.W. (1971). Influence of age on pain relief from analgesics. *Journal of the American Medical Association*, **217**, 1835–41.
36. Lewington, V.J., McEwan, A.J., Ackery, D.M., Bayly, R.J., Keeling, D.H., Macleod, P.M., Porter, A.T., and Zivanovie, M.A. (1991). A prospective randomised double-blind crossover study to examine the efficacy of strontiuim-89 in pain palliation in patients with advanced prostate cancer metastatic to bone. *European Journal of Cancer*, **27**, 954–8.
37. Bromley, D.B. (1974). The terminal stage: dying and death. *The psychology of human ageing*, pp. 267–87. Penguin, London.
38. Hahn, D.E.E. and van Dam, F.S.A.M. (1988). Psychosocial complications and implications of cancer treatment in the elderly. In *Cancer in the Elderly*, (ed. T. Bokhel), pp. 16–31. Excerpta Medica, Amsterdam.
39. Holland, J. and Massie, M. (1987). Psychosocial aspects of cancer in the elderly. *Clinics in Geriatric Medicine*, **3**, 533–9.
40. Cassileth, B.R., Lusk, E.J., Strouse, T.B., Miller, D.S., Brown, L.L., and

Cross, P.A. (1985). A psychological analysis of cancer patients and their next-of-kin. *Cancer*, **55**, 72–6.

41. Fancourt, G. and Castleden, M. (1986). The use of benzodiazepines with particular reference to the elderly. *British Journal of Hospice Medicine*, **35** (5), 321–6.

42. Jacoby, R.J. (1981). Depression in the elderly. *British Journal of Hospice Medicine*, **25** (1), 40–2.

43. Baldwin, B. (1989). An energetic approach to antidepressants in the elderly. *Geriatric Medicine*, **19** (3), 69–74.

44. Levine, P.M., Silberfarb, P.M., and Lipowski, Z.J. (1978). Mental disorders in cancer patients. *Cancer*, **42**, 1385–91.

45. Oxman, T.E. and Silberfarb, P.M. (1987). Psychiatric aspects of cancer in the aged. *Cancer Surveys*, **6**, 512–19.

46. Spiegel, D., Bloom, J.R., Kraemer, H.C., and Gottheil, E. (1989). Effect of psychosocial treatment on survival of patients with metastatic breast cancer. *Lancet*, **2**, 1209–10.

47. Bindemann, S., Soukop, M., and Kaye, S.B. (1991). Randomised controlled study of relaxation training. *European Journal of Cancer*, **27**, 170–4.

48. Kellett, J. (1989). The reality of sexual behaviour in old age : there's a lot of it about. *Geriatric Medicine*, 17–18.

49. Kellett, J. (1989). Sex and the elderly. *British Medical Journal* **299**, 934.

50. Neale, B. (1991). *Informal palliative care: a review of research on needs, standards and service evaluation*, Monograph No. 3. Trent Palliative Care Centre, Sheffield.

51. Cartwright, A. (1983). *Health surveys in practice and in potential.*

52. McCarthy, M. and Higginson, I. (1991). Clinical audit by a palliative care team. *Palliative Medicine*, **5**, 215–21.

53. Dickinson, D.J. and Ebrahim, S. (1990). Adding life to years: quality and the health care of the elderly. *Geriatric Medicine*, **20** (9), 11–17.

54. Beck-Friis, B., Strang, P., and Eklund, G. (1989). Physical dependence of cancer patients at home. *Palliative Medicine*, **3**, 281–6.

55. Mosley, J.G. (1988). Special problems for patients at the extremes of life. In *Palliation in malignant disease*, (ed. J.G. Mosley), pp. 139–45. Churchill Livingstone, Edinburgh.

56. Herd, E.B. (1990). Terminal care in a semi-rural area. *British Journal of General Practice*, **40**, 248–51.

57. Henderson, J., Goldacre, M.J., and Griffith, M. (1990). Hospital care for the elderly in the final year of life: a population based study. *British Medical Journal*, **301**, 17–19.

58. Dunlop, R.J., Davies, R.J., and Hockley, J.M. (1989). Preferred versus actual place of death: a hospital palliative care support team experience. *Palliative Medicine*, **3**, 197–201.

59. Townsend, J., Frank, A.O., Fermont, D., Dyer, S., Karran, O., Walgrove, A., and Piper, M. (1990). Terminal cancer care and patients' preference for place of death: a prospective study. *British Medical Journal*, **301**, 415–17.

60. Welch, T. Existential and spiritual concerns. In *Outpatient management for advanced cancer*, (ed. J.A. Billings), pp. 260–8. J.B. Lippincott, Philadelphia, PA.

61. O'Neill, Dugan, S., and McDowell Scallion, P. (1987). Nursing care of elderly persons throughout the cancer experience. *Clinics in Geriatric Medicine*, **3**, 517–31.

62. Reed, P.G. (1987). Spirituality among terminally ill hospitalised adults. *Research in Nursing and Health*; **10**, 335–44.
63. Thomas, K.J., Carr, J., Westlake, L., and Williams, B.T. (1991). Use of non-orthodox and conventional health care in Great Britain. *British Medical Journal*, **302**, 207–10.
64. Byrne, P., Dunstan, G.R., The Lord Jakobvits, Jayaneera, R.L.A., Marshall, J., Philipp, E.E., *et al.* (1991). Hospice care: Jewish reservations considered in a comparative ethical study. *Palliative Medicine*, **5**, 187–200.
65. Anonymous. (1989). Terminal careless. *British Medical Journal*, **299**, 1471. 1471.
66. Holli, K. and Hakama, M. (1978). Treatment of the terminal stages of breast cancer. *British Medical Journal*, **298**, 13–14.
67. Fottrell, E. (1990). Multidisciplinary functioning: will it still be of use? *British Journal of Hospice Medicine*, **43**, 253.
68. Monfardini, S. and Chabner, B. (1991). Joint NCI-EORTC consensus meeting on neoplasia in the elderly. *European Journal of Cancer*, **27**, 653–4.

Index